FOOD CHEMISTRY

FOOD CHEMISTRY

Alex V. Ramani
Reader
Department of Chemistry
St. Joseph's College (Autonomous)
Tiruchirappalli
Tamil Nadu

MJP PUBLISHERS

Cataloguing-in-Publication Data

Ramani, Alex V (1960 –   ).
Food Chemistry / by Alex V. Ramani. –
Chennai : MJP Publishers, 2009.
    xx, 300 p.; 21 cm.
Includes Glossary, references and index.
ISBN 978-81-8094-061-3 (pbk.)
    1. Chemistry, Food 2. Food Chemistry
I. Title.
    540:641.3 dc22 RAM MJP 058

ISBN 978-81-8094-061-3   **MJP PUBLISHERS**
© Publishers, 2009   47, Nallathambi Street
All rights reserved   Triplicane
Printed and bound in India   Chennai 600 005

Publisher : J.C. Pillai
Managing Editor : C. Sajeesh Kumar
Project Editor : P. Parvath Radha
Acquisitions Editor : C. Janarthanan
Editorial Team : B. Ramalakshmi, N. Pushpa Bharathi,
    L. Mohanapriya, Lissy John,
    N. Yamuna Devi, M. Gnanasoundari
CIP Data : Prof. K. Hariharan, Librarian
    RKM Vivekananda College, Chennai.

To

My beloved parents Mary and Vincent,
My wife Gracelin, and
My daughters Chrisie and Nessie

PREFACE

The Holy Bible says "Man doesn't live by bread alone; but, by every word of God". It is very much true. Man is destined to win bread by the grace of God. Food is the basic need for life. Without food, there is no life. This book is aimed at explaining the chemical aspects of food such as the basic chemical substances that make the food, the way the food can be analysed, etc.

Food helps in the well-being of man only if it has the proper amount of nutrients required for the biochemical processes. Food affects health if it has harmful substances that hamper biochemical reactions. The present-day foodstuff are adulterated for commercial purposes. The contaminants affect the health of the people who consume such foods. For example, the soft drinks have artificial flavours and colours of fruits. So, the drinks that we consume are not real fruit juices, but mere slow poisons. Analyses of such modern foodstuff reveal the fact that they are contaminated.

This book is intended to create an awareness in people about the diet pattern, healthy food, contaminants and the methods to identify the adulterants, etc.

I am happy to acknowledge the support rendered by Rev. Dr. T. Rajarathinam, S.J., Principal, St. Joseph's College (Autonomous), and the Former Principals Rev. Fr. A.G. Leonard, S.J., and Rev. Fr. S. John Britto, S.J., who have encouraged me in the writing of this book. I wish to thank Dr. N. Subramani, (Former Head, Department of Chemistry), Dr. X. Rosario Rajkumar, Head, Department of Chemistry, and all my colleagues who have contributed their valuable suggestions and ideas to prepare this book. I sincerely offer my thanks to Prof. Dr. K.V. Raman, Department of Chemistry, who has been a constant inspiration and guidance to me in bringing out this book. I thankfully remember my students who have worked elaborately to experiment the different analytical procedures.

I place on record the efforts of MJP Publishers and their editorial team for their keen interest in bringing out this book.

Alex V. Ramani

CONTENTS

1

FOOD, NUTRITION AND HEALTH

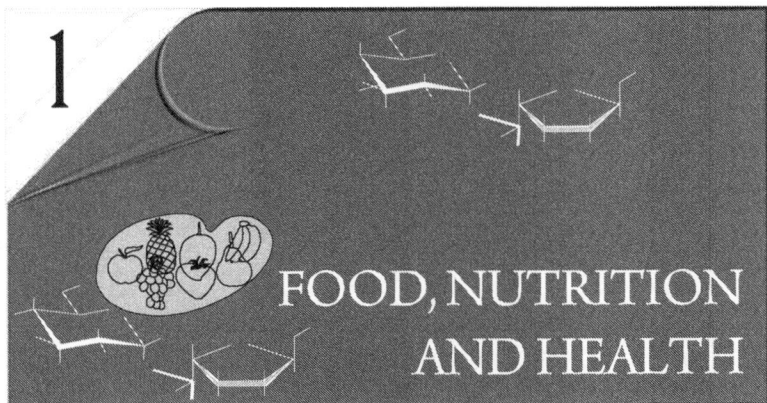

THE MEANING OF FOOD

Next to the air we breathe and the water we drink, food has been basic to our existence. It satisfies our hunger and has been the primary concern of mankind in its physical environment. The presence of food or the lack of it influences the destiny of humans. People eat to live. The ability to work, to lead a long and happy life, etc. are influenced by the food that one eats. Food gives an opportunity for fellowship with friends and family. Eating is a way to celebrate an event. Food is a symbol of prestige. Food is an enjoyment and leisure. Above all, food is a gift of God for which we must be thankful to Him.

Food supplies the body with energy-producing materials. Also, it provides substances for building and maintenance of the body. Food regulates the body processes. Thus food has many physiological functions to play.

WHAT IS HEALTH?

The World Health Organization (WHO) defines health as a state of complete physical, mental and social well-being. It is not merely the absence of disease or physical deformity.

NUTRIENTS

Nutrients are the basic constituents of food that must be supplied to the body in appropriate amounts. There are six different nutrients: carbohydrates, proteins, lipids, vitamins, minerals and water.

CLASSIFICATION OF FOOD

Food is classified based on the physiological functions. The foodstuff that contains larger proportion of nutrients like carbohydrates and lipids are called energy-yielding food. The food that contains greater proportions of proteins, minerals and water is called body-building food. The food that has higher amounts of proteins, minerals, vitamins and some fat is called body-regulating food (Table 1.1).

Table 1.1   Food, nutrients and their functions

Food	Nutrients	Functions
Cereals, roots, tubers, dried fruits, sugars, fats, etc.	Carbohydrates, lipids	Energy-yielding
Milk, meat, fish, egg, pulses, oil seeds, nuts, fish liver oil, etc.	Proteins, minerals, water	Body-building
Milk, egg, green leaves, fruits, etc.	Proteins, minerals, water, vitamins	Regulating

NUTRITIONAL STATUS AND CARE

Nutritional status is the condition of health of an individual influenced by the utilization of nutrients. It is determined by correlating the information on the medical and dietary history of individuals with medical examination reports. Nutritional care involves keeping the nutritional status at a higher level. Nutritional care can be provided through many ways—health education to people, a movement towards a better food management, dietary counselling, etc.

Malnutrition

Malnutrition is the impairment of health due to deficiency or excess of nutrients in food. There are two types of malnutrition. One is under-nutrition and the other is over-nutrition.

Under-nutrition is the deficiency of calories and/or deficiency of one or more nutrients. Over-nutrition is the excess of calories and/or excess of one or more nutrients.

Good nutrition is the efficient use of food within the home, public place and institution. It ensures good health. Good nutrition is achieved by the following measures:

* Production of sufficient amount of food by the application of Science and Technology.

* Preserving the food for a long time by the application of the best processing techniques.

* Efficient harvesting, transporting and storage techniques.

* Governmental measures like adequate checks and balances to ensure quality food.

* Educational programmes on nutrition for the students and others.

* Cultural programmes to create awareness on good nutrition among people.

Global Problems of Nutrition

Every year about 20 million people die of starvation. Most of them are children below the age of five. The reasons for these starvation deaths are poverty, ignorance, malnutrition, diseases, etc. More than one billion people are severely undernourished. Three-fourth of the world population lives in the Indian subcontinent, South Asia, and Africa. It is predicted that the global food crisis would worsen more than the energy crisis in another twenty years.

The world population keeps on increasing by about 0.2 million per day. The developing nations contribute a major share of it. These nations face difficulties in providing the required food supply, even though the food production is enhanced by modern technology. That is why these nations have opted for family planning measures. But, these measures do not work effectively for the following reasons:

    ❋   need for the children to sustain old parents

    ❋   child labour

    ❋   religious beliefs that family planning is a sin against God, and children are the gifts of God.

The energy crisis also affects the food supply in the world. Energy is the prime need for human survival. It is needed to produce fertilizers, to run machinery, to transport, etc.

Malnutrition is the greatest problem of nutrition throughout the world. Kwashiorkor, and marasmus, which affect the children, are the result of malnutrition. Vitamin deficiency leads to diseases like anaemia, blindness, rickets, scurvy, beriberi, pellagra, goitre, etc.

THE BODY COMPOSITION AND THE NUTRIENTS

The body is composed of water, proteins, fats, carbohydrates, minerals, and vitamins. Water, proteins and fats are in greater amount in the body. Carbohydrates are the important source of fuel, and are derived from food or formed from the fats and proteins in the living system itself. The approximate composition of the body is shown in Table 1.2.

Table 1.2 Nutritional composition of the body

Constituent	% Weight	Constituent	% Weight
Water	63	Minerals	
Protein	16	Ca	1.7
Fat	15	P	1.0
Carbohydrate	0.7	K	0.35
Vitamins	Very small amount	Na	0.17
		Mg, Zn, Fe	3.8

Amount of Nutrients in the Body

The nutritional processes of an organism are the sum of the physical and chemical changes that occur within the cell and the relationships that exist between cells and the surrounding environment. Cell biology, physiology and biochemistry are the sciences that describe these processes.

The major chemical constituents of the body are oxygen (65%), carbon (18%), hydrogen (10%), nitrogen (3%) and other elements (4%). These elements are in the form of water, protein, fat, glycogen, etc., and contribute to 96% of the body weight. The remaining are minerals that are 4%. Water, present in all the body tissues, accounts for 55 to 70% of the body weight. The water content is inversely related to the fat content. Normal individuals have high water content and low fat content. Obese persons have higher fat content and lower water content. A steady increase in body fat content is observed while aging. Protein accounts for 18% of the body weight. Only 300 g of sugar is present in the body, in the form of glycogen. Most of the vitamins, hormones and enzymes are present in the body only in very low levels and they do not contribute significantly to the body weight.

FUNCTIONS OF FOOD

The nutrients derived from food, supply energy for the activity of the body. They provide many structural materials and many more regulatory substances. Sugars and fats are sources of energy. Proteins, minerals, vitamins and water aid growth and regulatory activities.

METABOLISM

Metabolism involves release of nutrients from food, transformation of the nutrients required for the energy release, syntheses of regulatory materials, etc. It consists of two processes: one is an anabolic (constructive) process and the other is a catabolic (destructive) process.

Anabolism is a constructive process in which new substances are produced from simple compounds that are obtained from the food. For example,

Amino acids ⟶ Enzymes, hormones, proteins

Glucose ⟶ Glycogen, other sugars

Fatty acids ⟶ Cholesterol, other sterols

Catabolism is a destructive process in which the complex substances are broken down into simpler compounds. For example,

Proteins $\longrightarrow$ Amino acids

Glucose $\longrightarrow$ Carbon dioxide, water, heat

Fats $\longrightarrow$ Glycerol, fatty acids, etc.

The metabolic pathway of a single nutrient may be traced in order to study the metabolism of that particular nutrient. For example, the carbohydrate is tracked down during digestion, absorption and intermediary metabolism for understanding the metabolism of carbohydrates. But, there is a multitude of metabolic events of other nutrients interlinked with the metabolic pathway of a single nutrient. There is always a balance between anabolism and catabolism. When anabolism is more than the catabolism, it reflects growth. Loss of body weight and loss of essential substances signify that catabolism is more than anabolism.

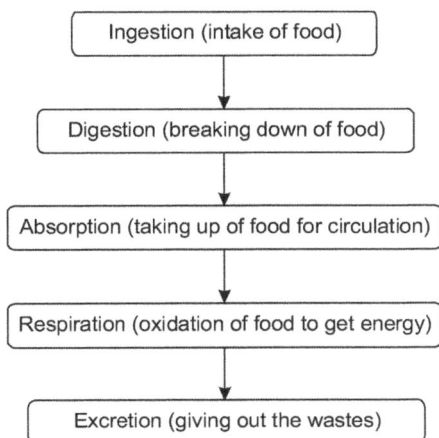

Ingestion (intake of food)
↓
Digestion (breaking down of food)
↓
Absorption (taking up of food for circulation)
↓
Respiration (oxidation of food to get energy)
↓
Excretion (giving out the wastes)

As shown in the flowchart, metabolism comprises the sequence of events starting from ingestion up to excretion. The metabolic processes occur in different organs in a living species.

CELL AS A FUNCTIONAL UNIT

The cell is a microscopic structure that is the most fundamental unit of any living organism. It is just like the brick that forms the fundamental unit of a building. The survival of a cell depends on its favourable environment. In simple living things, cells have fewer functions. But, in complex living things, cells have multiple functions. Figure 1.1 depicts a cell and its parts.

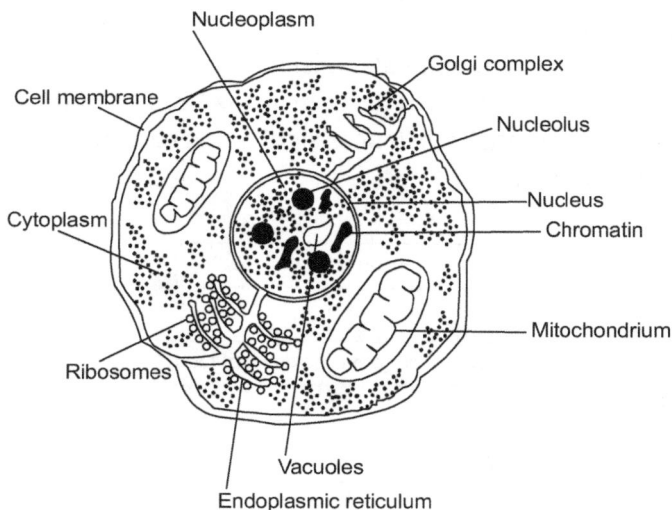

Figure 1.1   Cell structure

Cell membrane   This surrounds the cytoplasm. It maintains the internal environment of the cell membrane.

Nucleus   Nucleus is the central part of a cell. It is the source of DNA, the genetic plan for the construction of proteins.

Cytoplasm   This fills the space between the nucleus and the cell wall. It surrounds the organelles of a cell.

Mitochondrion   It is a rod-shaped structure that contains hundreds of oxidative enzymes. It is the powerhouse of a cell.

Lysosomes   These are the bags inside a cell that contain digestive enzymes.

Endoplasmic reticulum   It is a channel that allows the flow of materials to and from the various parts of the cell to the extracellular environment.

Golgi bodies   These store and concentrate the enzymes and release them on demand.

DIGESTION

It is the process by which complex food materials are broken down into simple chemical substances that are suitable in size and composition for absorption. It consists of a series of physical and chemical changes effected on the food. There are two kinds of digestion: one is physical digestion and the other chemical digestion.

Physical digestion consists of grinding, crushing and mixing of food with digestive juices and carrying the food through the digestive tract by propulsive motion. Chemical digestion involves breaking down of the food by digestive juices and enzymes (endo- and exoenzymes).

All the physical and chemical changes are regulated through neural and hormonal mechanisms. Hormones such as gastrin, secretin and cholecystokinin are involved in hormonal regulation. Table 1.3 gives the important functions of gastrointestinal hormones.

Digestion in the Mouth

Teeth grind and crush the food into small particles. The crushed food mass is mixed with saliva (three pairs of salivary glands produce about 1.5 L saliva daily). Starch is digested by α-amylase (ptyalin). Then, the food mass is pushed into the stomach.

Digestion in the Stomach

The food mass is mixed with gastric secretions by the wavelike contractile motion of stomach muscles. In this process of gastric digestion, the food becomes a semi-solid. Major chemical digestion takes place only in the stomach. The gastric juice secreted is 2–2.5 litres per day. First, carbohydrates present in food are digested. Then, proteins and finally fats are digested. The stomach is emptied of food within 4 hours.

Digestion in the Intestine

The small intestine plays a very active role in the digestion. It is as long as 22 feet. It has three sections—duodenum, jejunum and ileum. The food is digested chemically by duodenal, bile and pancreatic juices.The gastrointestinal hormones and their functions are listed in Table 1.3.

ABSORPTION

It is a complex process (Figure 1.2) by which the simple and important chemical substances of the food are taken up.

Table 1.3   Important functions of gastrointestinal hormones

Hormone	Site of release	Stimulant	Organs affected	Effect on organs
Gastrin	Antral mucosa of stomach, duodenum, jejunum	Polypeptidase, amino acid, caffeine, alcohol, etc.	Oesophagus, stomach	Pressure increase, HCl secretion, pepsinogen secretion, gastric antral motility
Secretin	Duodenal mucosa	Gut acidity (pH 4–5)	Oesophagus, stomach, duodenum, pancreas, liver	Controls pressure increase, gastrin and duodenal motility reduction, increased output of water, increased output of electrolyte
Cholecystokinin Pancreozymin	Duodenal mucosa	Amino acids, HCl, fatty acids	Stomach, gall bladder, pancreas	Inhibition of gastrin, stimulation of secretion of acid, contraction of gall bladder, stimulation of enzyme secretion

The nutrients are dissolved in the lipid of the cell membrane and then diffused into the cells. There are four types of absorption mechanisms:

    i.   Absorption by pores

    ii.   Absorption by carriers

    iii.   Absorption by pumps

    iv.   Absorption by pinocytosis.

Mostly, the small and large intestines absorb the digested food.

Absorption by pores   The simple chemical substances are taken up into the cells through membranes that are made up of lipoproteins. The membranes have thousands of pores through which the electrolytes and other water-soluble molecules enter the cells.

Absorption by carriers   The ferrylike chemical substances shuttle back and forth through the cell membrane. These chemical ferries transport large fat-soluble substances such as amino acids, fats, etc.

Absorption by pumps   Some molecules require energy to move from the intestinal lumen into the mucosal cell. This requires pumping action. With the cellular energy or ATP and a carrier, substances like glucose, galactose and minerals, Na^+, K^+, Mg^{2+}, PO_4^{3-}, I^-, Ca^{2+} and Fe^{2+} are absorbed.

Absorption by pinocytosis   It is a "drinking-in" process. Large molecules are absorbed by pinocytosis.

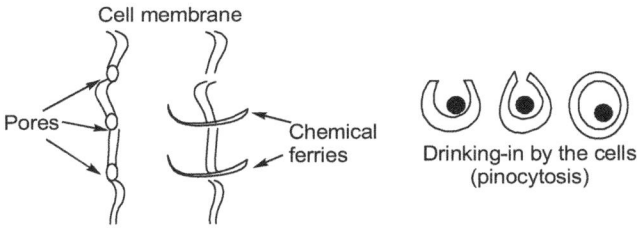

Figure 1.2   Absorption mechanism

INTERMEDIARY METABOLISM

Intermediary metabolism refers to the physical and chemical changes that take place in the internal environment. These activities occur within each cell. The nutrients diffuse from arterial capillary blood into the interstitial fluids surrounding the cells. They are then absorbed. Similarly, the cells dispose the waste substances to the interstitial fluid and in turn to the venous circulation. Figure 1.3 depicts the chemistry of various secretions in digestion and absorption of food.

The enzymes of a given cell determine the specific functioning of a cell. For example, glucose is used up by the cell by the action of various enzymes one after the other in an assembly-line fashion until the desired end product is formed. Even if a single enzyme is missing, there is a breakdown in the assembly line, and problems arise.

A number of nutrients are available in the cell at any moment for the metabolic purposes. For example, glucose, fatty acids, glycerol and amino acids can enter the common pathway that yields energy. Sometimes glucose can be metabolized to fatty acids and cholesterol. Likewise, amino acids can act as the source of glucose and fatty acids and so on. The scheme of biochemical changes in the intermediary metabolism is shown Figure 1.4.

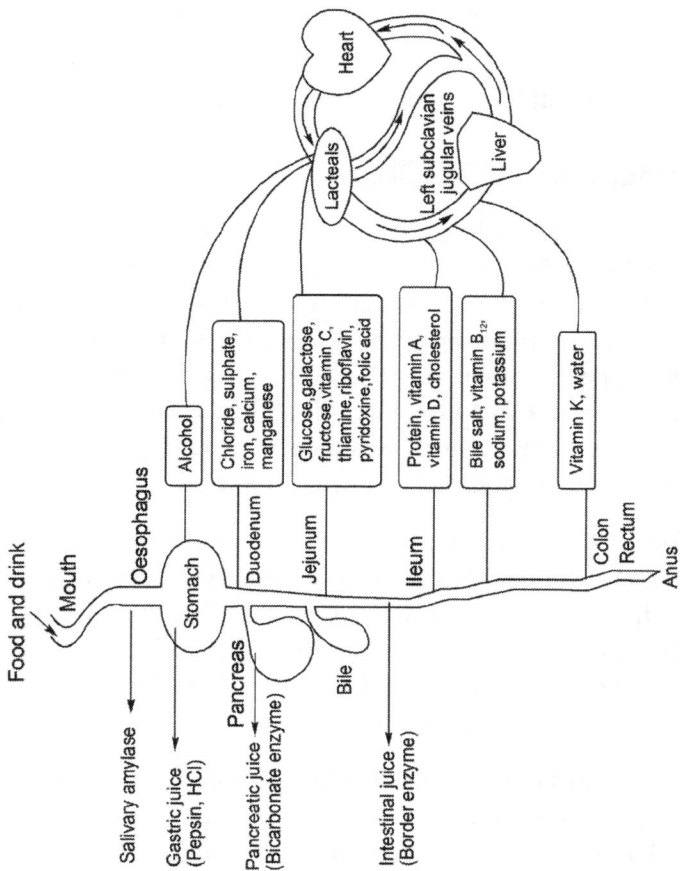

Figure 1.3   Chemistry of various secretions in digestion and absorption of food

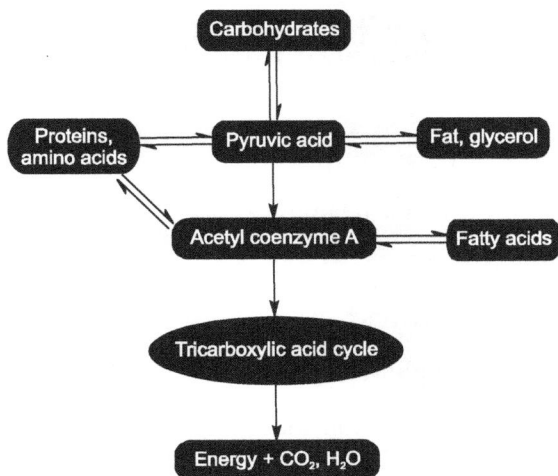

Figure 1.4   Chemistry of intermediary metabolism

THE SCIENCE OF NUTRITION

Nutrition is a multi-disciplinary science. Figure 1.5 depicts the interlinks between the different fields of nutrition.

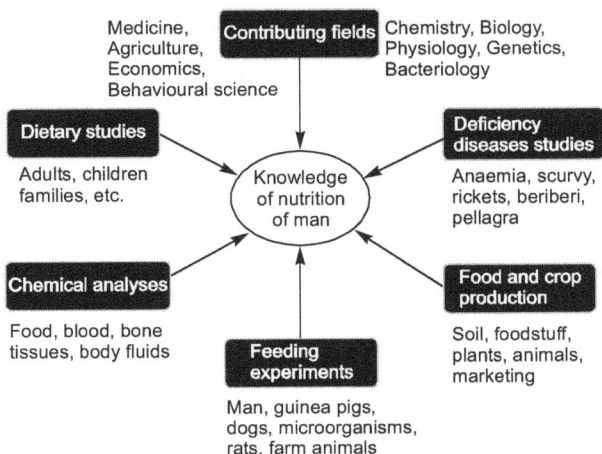

Figure 1.5   Interlinks between the different fields of nutrition

REVIEW QUESTIONS

Give short answers

1. List the different meanings of food.

2. Define WHO.

3. What are nutrients?

4. What are the physiological functions of food?

5. Write the nutritional classification of food, with two examples for each.

6. What does "good health" signify?

7. What are micronutrients?

8. What is malnutrition?

9. Define under-nutrition.

10. Explain metabolism.

11. What is the sequence of events during metabolism?

12. How are the nutrients, proteins, sugars and lipids chemically transformed in catabolism?

13. Define digestion.

14. Name the digestive juices that are secreted in the intestine.

15. List the four types of absorption mechanisms.

16. List the major chemical constituents of the human body.

17. Over-nutrition means _____ and/or _____.

18. Sugars and lipids are the important sources of _____.

19. The nutrients that are essential for the regulatory activities are _____.

20. Starch is digested by the enzyme _____ in the mouth.

Give detailed answers

1. Discuss the global problems of nutrition.

2. Give the body composition based on the following constituents: water, protein, fat, carbohydrate, minerals.

3. Show that nutrition is a multi-disciplinary field.

4. Draw the structure of the animal cell, indicating its various parts and their functions.

5. Discuss the digestion processes in the mouth, stomach and intestine.

6. Explain the four types of food absorption mechanisms.

7. Write a summary of secretion and absorption.

8. Match the following.

1. Pinocytosis	a. constructive process
2. Ptyalin	b. "powerhouse" of the cell
3. Mitochondrion	c. lipid
4. Cholesterol	d. alpha-amylase
5. Metabolism	e. drinking-in process

2

CARBOHYDRATES

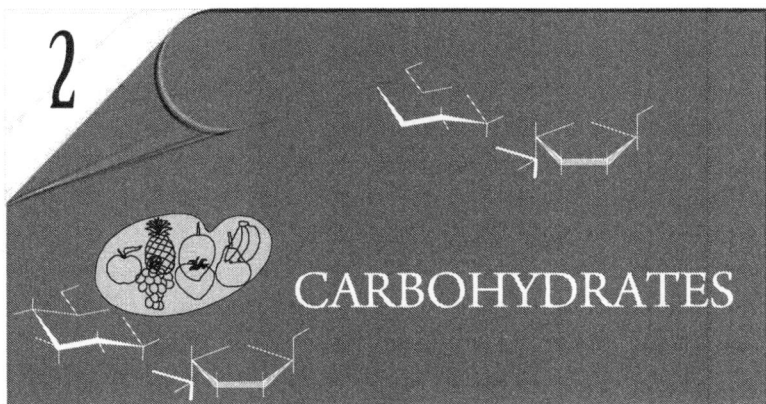

INTRODUCTION

Carbohydrates are one of the nutrients present in food. They have the general formula $C_n(H_2O)_n$ where n can be in the range of 3 to 1000. Carbohydrates may be defined as polyhydroxy aldehydes or ketones and/or their derivatives. They are present in fruits and vegetables. These sugars are found in seeds, roots and tubers; they form the constituents of structural tissues.

Carbohydrates are the most economical energy source of food. Carbohydrates are required by the body to use fat efficiently. Some animals have the exoskeleton made of glucosamine polymer.

The presence of excess of unabsorbed sugar in the human body leads to the physiological disorder called "diabetes". Some industries like milling, baking, brewing, syrup and sugar are based on carbohydrates.

CLASSIFICATION OF CARBOHYDRATES

Carbohydrates are classified into two broad classes—simple sugars (non-hydrolysable) and compound sugars (hydrolysable).

Simple sugars are otherwise called monosaccharides and compound sugars are called di, tri, oligo, and polysaccharides. Table 2.1 gives the list of different types of sugars. It is conventionally understood that mono- and disaccharides are sugars and polysaccharides are non-sugars. There is another classification of sugars based on the reactions with Fehling's and Tollen's reagents. One class of sugars is termed reducing sugars as these react with Fehling's and Tollen's reagents. Another class of sugars is the non-reducing sugars, which do not react with Fehling's and Tollen's reagents.

Table 2.1   Classification of carbohydrates and their nutritional sources

Carbohydrates		Nutritional sources
Type	**Examples**	
Monosaccharides	Glucose	Fruits, vegetables, corn sugars, molasses
	Fructose	Nectar, fruits, vegetables
	Galactose	Not freely present in nature
Disaccharides	Sucrose (fructose + glucose)	Fruits, vegetables, sugar cane
	Maltose (glucose + glucose)	Sprouting grains, cereals, milk
	Lactose (glucose + galactose)	Milk (mostly animal source)
Polysaccharides	Starch	Roots, seeds, tubers
	Dextrin	Partially cooked food
	Glycogen	Liver, muscle of animals
	Cellulose	Stem and leaves of vegetables, seed and grain coverings, skins and hulls
Mucopolysaccharides	Hyaluronic acids, chondroitin, heparin	Produced during some biochemical processes in living things

Monosaccharides are named aldoses (aldehyde functionality) and ketoses (ketone functionality). For example, the simple sugars are named as aldotriose, ketotriose, etc.

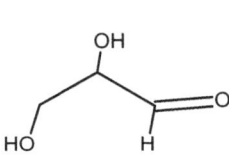

Aldotriose

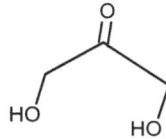

Ketotriose

Some sugars are given names based on the straight chain structure and ring structure, e.g. aldohexose, ketohexose, etc.

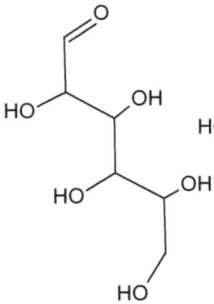

Aldohexose

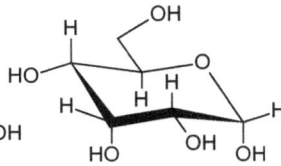

α-D-Allopyranose

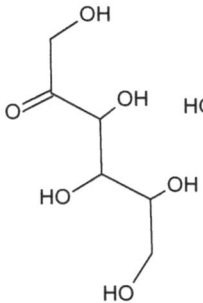

Ketohexose

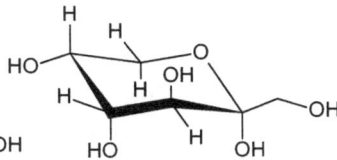

α-D-Fructopyranose

GENERAL PROPERTIES

All mono- and disaccharides are colourless solids. They are soluble in water and sparingly soluble in alcohol. They are sweet in taste. The relative sweetness of some mono- and disaccharides are given below:

Fructose	110–175
Sucrose	100
Glucose	75
Galactose	35–70
Lactose	15–30

Simple sugars can form open-chain and cyclic structures.

Physical and Chemical Properties of Mono- and Disaccharides

Glucose is a colourless crystalline solid. It melts at 146°C. It is readily soluble in water and sparingly soluble in alcohol. It is insoluble in less polar or non-polar solvents. It is dextrorotatory.

Reactions of Glucose

The chemical reactions of glucose are given in Figure 2.1. Glucose reacts with acetic anhydride to give the pentaacetyl derivative of glucose. With methanol and hydrochloric acid, it forms acetal. On fermentation by the microorganism, yeast, glucose yields ethanol and carbon dioxide. Hydroxylamine hydrochloride reacts with glucose to yield an oxime derivative. Glucose on reduction by hydrogen and nickel (catalyst) gives glucitol. Dilute nitric acid oxidizes glucose to glucaric acid. The reagents, bromine-water, Fehling's solution or Tollen's reagent oxidize glucose to gluconic acid. Glucose undergoes condensation with phenylhydrazine to form hydrozone derivative.

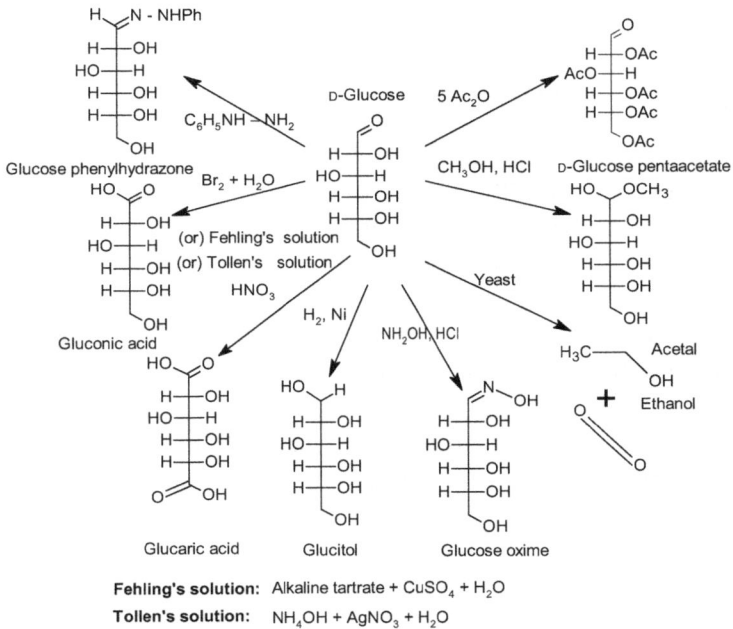

Fehling's solution: Alkaline tartrate + $CuSO_4$ + H_2O

Tollen's solution: NH_4OH + $AgNO_3$ + H_2O

Figure 2.1   Chemistry of glucose

Mutarotation   The optical rotation of glucose changes when it is kept in solution form. It is due to the interconversion between two forms (α and β-forms). The mechanism is shown below:

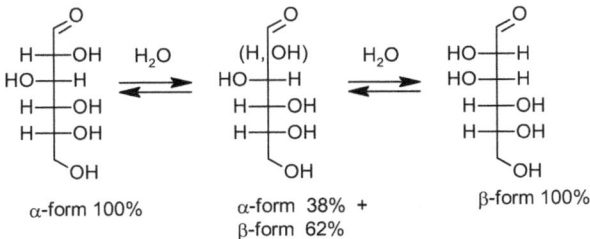

Fructose is a colourless crystalline solid. It melts at 102°C. It is readily soluble in water and sparingly soluble in alcohol. It is insoluble in less polar or non-polar solvents. It is laevorotatory in nature.

Reactions of Fructose

The chemical reactions of fructose are given in Figure 2.2. Fructose reacts with acetic anhydride to give the pentaacetyl derivative of fructose. With methanol and hydrochloric acid it forms acetal. Fructose on reduction by sodium borohydride gives glucitol. On fermentation by the microorganism, yeast, fructose yields ethanol and carbon dioxide. Dilute nitric acid oxidizes fructose to trihydroxyglutaric acid and formic acid. The reagent bromine-water oxidizes fructose to form tartaric acid, glycolic acid and formic acid. Glucose undergoes condensation with phenylhydrazine to form the hydrozone derivative.

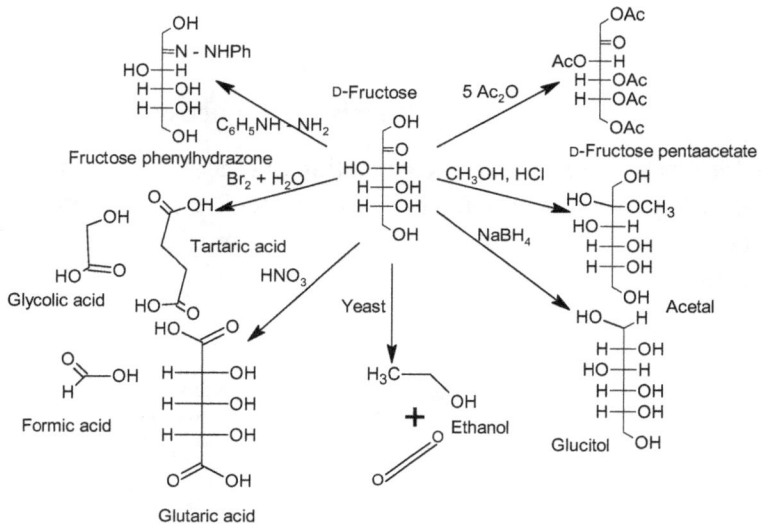

Figure 2.2   Chemistry of fructose *(Continues)*

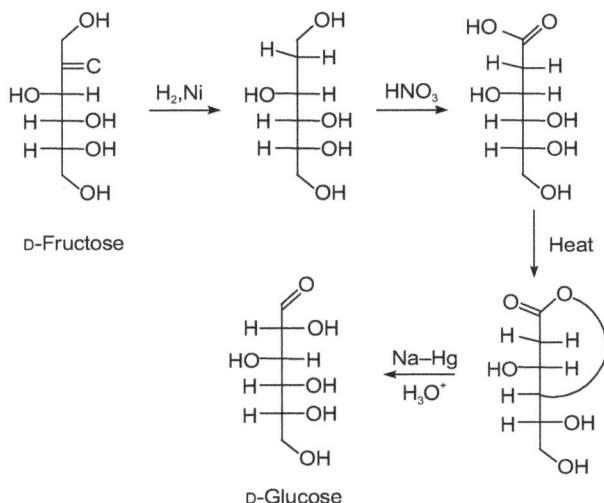

Figure 2.2   Chemistry of fructose

Fructose is converted to glucose by a sequence of reactions. Fructose is reduced by hydrogen and nickel catalyst to give glucitol and it is oxidized by dilute nitric acid to form glucaric acid. This acid is heated to form a lactone type of compound which is further reduced by sodium amalgam in water to yield glucose.

Reactions of Sucrose

The chemical reactions of sucrose is illustrated in Figure 2.3. Change of optical rotation from positive to negative is called inversion of sugar. It occurs when sucrose solution is allowed to stand for sometime in a slightly acidic condition. Sucrose reacts with acetic anhydride to give octa-acetyl derivative. Sucrose undergoes fermentation reaction with yeast to give ethanol. It is an age-old practice of preparing alcoholic beverage. Sucrose reacts with hydrochloric acid and gives laevulinic acid.

Sucrose shows charring reaction while heating with sulphuric acid, thereby carbon is formed ultimately. With nitric acid, sucrose gives oxalic acid.

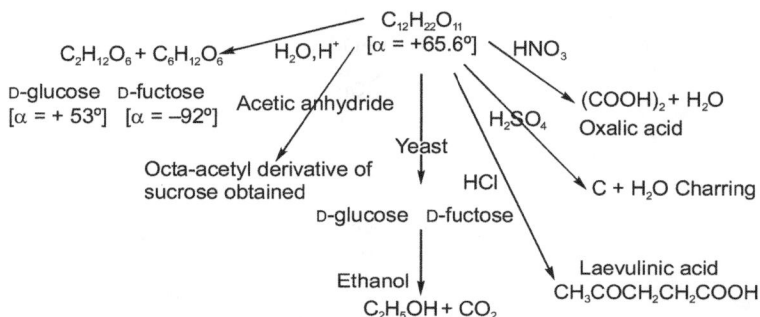

Figure 2.3   Chemistry of sucrose

Note   Sucrose is esterified along with fatty acid at any one of the hydroxyl functionality, to produce a non-ionic, biodegradable detergent. It has utility value.

Ring Structures of Glucose, Fructose and Sucrose

All sugars have pyranose or furanose ring structures. Because of the ring structures they show optical isomers less than the number that are predicted for the open-chain structures. The ring structures for some sugars are given in Figure 2.4. The alpha- and beta-labelling of glucose and fructose arise due to change in the spatial orientation of H and OH at the C1 carbon of the ring. This is called anomerism.

MANUFACTURE OF CANE SUGAR

Cane sugar is the commercial name of sucrose. We use it in our day-to-day life. The principle of manufacture of cane sugar is briefly discussed.

Sugar cane is crushed to get its juice. The crushed waste (bagasse) is used as fuel. The juice is purified using milk of lime and sulphur dioxide or carbon dioxide. The juice is then filtered to remove suspended impurities through activated charcoal. The clear filtrate is concentrated by multiple evaporators. A syrupy liquid is obtained. It is boiled in vacuum till sugar crystals appear. These crystals are again centrifuged in order to remove the mother liquor (molasses). The crystals are sprayed with little water and again centrifuged to remove the mother liquor. Thus, pure sugar crystals are obtained.

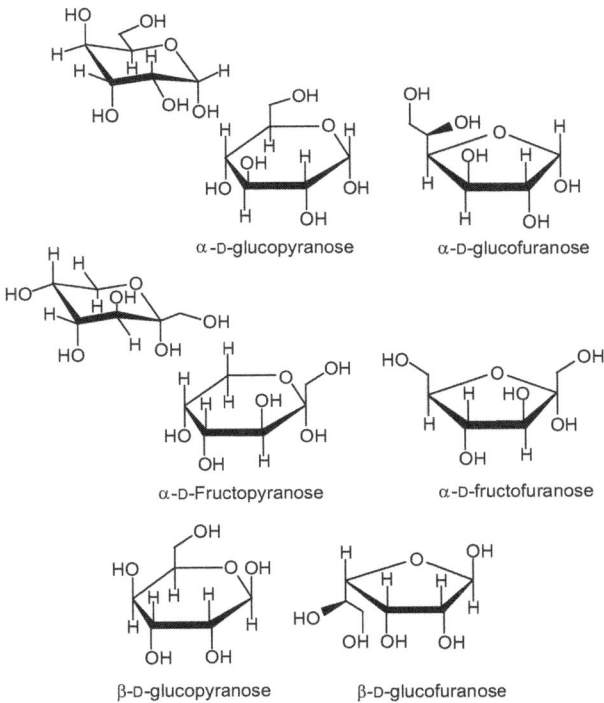

α-D-glucopyranose   α-D-glucofuranose

α-D-Fructopyranose   α-D-fructofuranose

β-D-glucopyranose   β-D-glucofuranose

Figure 2.4 Ring structures of glucose, fructose and sucrose (*Continues*)

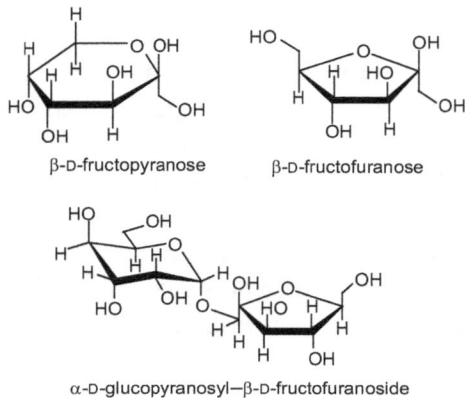

β-D-fructopyranose   β-D-fructofuranose

α-D-glucopyranosyl—β-D-fructofuranoside

Figure 2.4 Ring structures of glucose, fructose and sucrose

PHYSICAL AND CHEMICAL PROPERTIES OF POLYSACCHARIDES

Starch is a polysaccharide. It is a non-reducing sugar. Most of the vegetables contain this non-reducing sugar. It is a polymer of glucose and contains two polymeric chains—amylose and amylopectin.

Starch granules are insoluble in cold water. These granules swell in cold water. On boiling the water, the granules burst and form a colloidal solution that turns to a gel on cooling. When starch is acid-hydrolysed, it gives a mixture of low-molecular weight polysaccharides—dextrins and maltose. If the hydrolysis is continued, glucose is formed finally.

Amylose is the water-soluble part that is about 20 per cent, and amylopectin is the water-insoluble part that is about 80 per cent of the molecular structure. Amylose is a straight-chain polymer of glucose and amylopectin is a cross-linked polymer of glucose.

Starch is a polymer of glucose, which when hydrolysed totally with strong acid yields a number of glucose molecules. When the starch is partially hydrolysed, a number of maltose sugars are produced. Figure 2.5 shows some chemical reactions of starch.

Note Amylose gives the first two types of methylated products. Amylopectin gives three kinds of products, including 2,3-di-O-methyl D-glucose. This dimethylated product is a proof for the cross-links that exist in amylopectin.

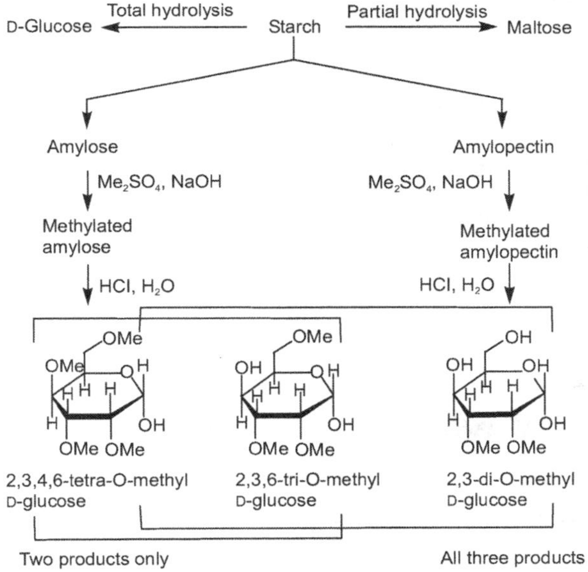

Figure 2.5 Chemistry of starch

Starch is a polysaccharide made of glucose. It is much useful in the manufacture of glucose and in the large-scale preparation of alcohol. It is used in paper, textile and printing industries.

MANUFACTURE OF STARCH

Starch is an important commercial food product. It is a polysaccharide and it is very much used by people in day-to-day life. The manufacture of starch is discussed very briefly (Figure 2.6).

The corn is soaked with water in a tank. Then, the corn is crushed and allowed to stand for sometime. The starch suspension is filtered through the sieve. The undesired solid mass is removed. The milky starch filtrate is collected and steam heated in a special chamber. Thus, the dry starch powder is obtained.

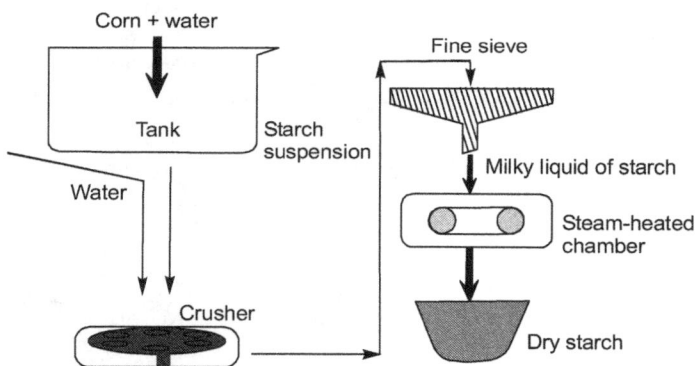

Figure 2.6   Flowchart of manufacture of starch

Cellulose   It is a form of polysaccharide that is resistant to the digestive enzymes in human beings. It remains in the digestive tract and contributes important bulk to the diet. This bulk helps in the movement of the food mass in the digestive tract and stimulates peristalsis. It forms the supporting framework of plants in the stems, leaves, seeds, grain coverings, skin, hull, etc. Cellulose is a colourless fibrous substance. It is insoluble in water. It gets dissolved in a solution of copper (II) hydroxide and ammonia [Schwetzer's reagent].

Dextrins They are polysaccharides and are the by-products of the reaction where starch is partially hydrolysed to form maltose. Dextrins form a soluble, gummy carbohydrate, which is used commercially as mucilage for envelopes and postage stamps or as sizing for adhesive tapes. Dextri-maltose is an infant formula preparation.

Pectin It is a non-digestible polysaccharide, existing in the colloidal form. It is found in fruits. It possesses a thickening quality and is used as a base for fruit jellies. It is used in cosmetics and drugs also.

Inulin It is a polysaccharide composed of fructose, and has little dietary value. It is commonly found in onion, garlic and artichokes. It is digested in the large intestine by bacteria and is stored in the unavailable form, immediately after digestion. But, after some time it is converted to available sugar. But, it is useful in the medical field to test the renal function. Inulin is filtered at the glomerulus to measure the filtration rate of the glomerulus. It is called **inulin clearance test**.

FUNCTIONS OF CARBOHYDRATES IN THE BODY

Energy source Carbohydrates are important energy sources. Body tissues require a constant dietary supply of carbohydrates. The amount of carbohydrate in the body of a man weighing 70 kg is about 365 g. This amount is stored in different forms as shown below:

Liver glycogen	100 g
Muscle glycogen	245 g
Extracellular blood sugar	10 g

This amount of sugar (365 g) is sufficient to provide energy for about thirteen hours of very moderate activity. Carbohydrates must be ingested regularly in frequent intervals to meet the energy demands of the body.

Protective action   In the liver, carbohydrates are not only oxidized as fuel, but they also detoxify the toxic substances (drugs). For example, glucuronic acid (a derivative of glucose) combines with toxic materials (drugs) to produce harmless forms that can be excreted.

Sparing action of protein   Carbohydrates have a regulating influence on protein metabolism. The presence of sufficient amount of carbohydrates for the energy need, prevents the sparing of protein for this purpose. Thus, a major portion of proteins are utilized for the basic structural purpose of tissue building.

Antiketogenic effect   The amount of carbohydrates consumed determines the breaking down of fat. Therefore, it affects the formation and disposal rates of ketones. Ketones are the intermediate products of fat metabolism. These ketones are further converted to fatty acids. When this conversion is incomplete, more ketone bodies get accumulated. This condition is called **ketosis,** which causes great damage to kidney as ketone bodies like acetoacetic acid, acetone and β-hydroxybutyric acid are to be removed in the excretion process. So, the presence of sufficient amount of carbohydrates prevents the formation and accumulation of excess ketone.

Maintenance of heart action   Heart action is a life-sustaining muscular exercise. The glycogen in cardiac muscle is an important emergency source of contractile energy. In a defective heart, glycogen availability is poor. It may cause cardiac symptoms or angina.

Maintenance of central nervous system function
A constant amount of carbohydrates is necessary for the proper functioning of the central nervous system, especially of the brain. The brain is maintained by a minute-to-minute supply of glucose through blood. Prolonged hypoglycaemic shock may damage the brain, irreversibly. So, carbohydrates are essential for the functional integrity of the nervous system.

DIGESTION OF CARBOHYDRATES

The first stage of digestion of carbohydrates occurs in the mouth when food is chewed. The saliva contains an enzyme, ptyalin (alpha-amylase), which splits starch into dextrin and maltose (at pH 7.0). The broken down starch fragments reach the stomach where they are mixed with acidic gastric juice. Now, the amylase activity is inhibited.

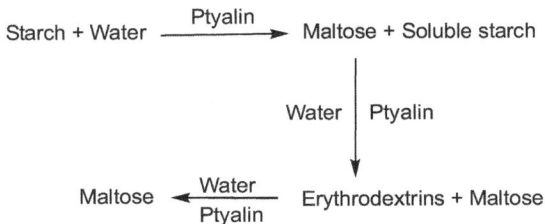

$$\text{Starch + Water} \xrightarrow{\text{Ptyalin}} \text{Maltose + Soluble starch}$$

$$\downarrow \text{Water} \quad | \quad \text{Ptyalin}$$

$$\text{Maltose} \xleftarrow[\text{Ptyalin}]{\text{Water}} \text{Erythrodextrins + Maltose}$$

Chemical digestion of carbohydrates takes place in the small intestines. There, many enzymes like pancreatic amylase, intestinal amylase, sucrase, lactase, maltase and isomaltase act on carbohydrates. Thus, glucose, fructose and galactose are obtained as the end products (Table 2.2).

Table 2.2   Carbohydrate digestion

Organ	Enzyme	Action
Mouth	Ptyalin	Starch → Dextrin → Maltose
Stomach	Ptyalin	Above action continued to a minor extent
Small intestine	Pancreatic amylase	Starch → Dextrin → Maltose
	Intestinal sucrose	Sucrose → Glucose + Fructose
	Intestinal lactase	Lactose → Glucose + Galactose
	Intestinal maltase	Maltose → Glucose + Glucose

ABSORPTION OF CARBOHYDRATES

Carbohydrates are absorbed into the bloodstream as glucose, galactose and fructose. The absorbing surface area of the small intestine is greatly enlarged by millions of villi (tiny, fingerlike projections of the mucous membrane). About 90% of the digested food materials are absorbed in the small intestine. Water and a little portion of the digested food are absorbed in the large intestine.

By the capillary action of the villi, simple sugars enter the portal circulation and are transported to the liver. Here, fructose and galactose are converted to glucose. The glucose is converted to glycogen and stored.

METABOLISM OF CARBOHYDRATES

The end product of carbohydrate digestion is glucose. Glucose is circulated through the blood to various parts of the body. The liver is the major site storage of glucose as glycogen. It handles glucose and much of the chemical activity takes place there. Other tissues like fat (adipose) tissue, muscle tissue and renal tissue play an important role in storing the glucose and in its chemical activity. Energy metabolism takes place in almost all cells. The mechanism of oxidation of glucose is expressed as follows:

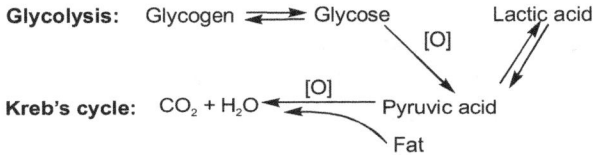

Glycolysis: Glycogen ⇌ Glycose   Lactic acid

                              [O]

                             [O]

Kreb's cycle: $CO_2 + H_2O$ ⟵ Pyruvic acid

                                  Fat

BIOSYNTHESIS OF SUGARS

The biosynthesis of the main α-glucan chain of starch amylose or glycogen amylopectin is similar to the synthesis of cellulose. The enzymes involved are **starch synthase** and **glycogen synthase**. They are generally called **α-glucan synthase**.

UDP-glucose and ADP-glucose are the substrates for the synthesis of starch and glycogen. The reaction is as follows:

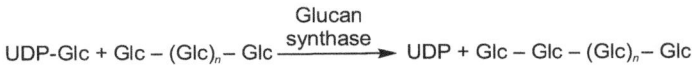

$$\text{UDP-Glc} + \text{Glc} - (\text{Glc})_n - \text{Glc} \xrightarrow{\text{Glucan synthase}} \text{UDP} + \text{Glc} - \text{Glc} - (\text{Glc})_n - \text{Glc}$$

The formation of α-(1-6) branch in amylopectin occurs in an interesting way. 1-4 and 1-6 α-branching enzyme cleaves a short oligonucleotide fragment from the non-reducing end of an α-glucan chain. Then, the enzyme transfers the fragment to C_6 hydroxyl part. The non-reducing end of this branch fragment can be further extended by the action of α-glucan synthase. That is how the sugars are biosynthesized.

ANALYSIS OF CARBOHYDRATES

Analysis of carbohydrates can be both qualitative and quantitative. In the qualitative analysis, the specific sugars are tested chemically for their presence or absence. In the quantitative analysis, the exact amounts of different sugars are determined by chemical methods.

Qualitative Analysis

Molisch test   2 mL carbohydrate solution + 2–3 drops of Molisch reagent. Warm. Mix well. Add 5 mL conc. sulphuric acid through the sides of the test tube. Formation of a dark violet ring indicates the presence of carbohydrate.

Molisch reagent   10 g alpha-naphthol + 100 mL alcohol.

Benedict test   2 mL carbohydrate solution + 4 mL of Benedict's solution. Mix well. Heat and cool. Yellow or red precipitate confirms the carbohydrate.

Benedict reagent

   Solution A: Sodium citrate + sodium carbonate in water.

   Solution B: Copper sulphate in water.

Seliwanoff test   2 mL carbohydrate solution + 2 mL of Seliwanoff's reagent. Boil in a water bath. Cool. Red precipitate indicates the presence of carbohydrate.

Seliwanoff's reagent   Resorcinol + HCl in water.

Iodine test   2 mL carbohydrate + 2 mL iodine reagent. Blue colouration signifies the presence of carbohydrate.

Iodine reagent   KI and iodine in water.

Bial's test   2 mL carbohydrate solution + 2 mL of the Bial's reagent. Boil and cool. Carbohydrate shows green colouration.

Bial's reagent   Orcinol + HCl + iron (III) chloride in water.

Quantitative Analysis

Reduction method This method is applicable to the reducing sugars. A definite amount of the sugar solution is mixed with a definite amount of Fehling's reagent, heated and cooled. The $-C=O$ group of the sugar gets oxidized by reducing the copper (II) sulphate to copper (I) oxide.

The copper (I) oxide precipitate is quantitatively separated by filtration. Then, it is dissolved in excess amount of ferric alum. The iron (III) ion in ferric alum is reduced to iron (II) ion. The iron (II) ion is quantitatively titrated with a standard potassium permanganate solution. Thus, the amount of copper (I) oxide formed and in turn the amount of sugar, is determined using the table of values.

The principal equation is given as follows:

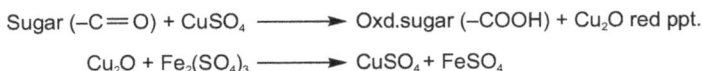

$$\text{Sugar } (-C=O) + CuSO_4 \longrightarrow \text{Oxd.sugar } (-COOH) + Cu_2O \text{ red ppt.}$$

$$Cu_2O + Fe_2(SO_4)_3 \longrightarrow CuSO_4 + FeSO_4$$

Hydrolysis method Mostly, polysaccharides are determined quantitatively by this method. Starch is subjected to enzyme hydrolysis.

The glucose formed is treated with 3,5-dinitrosalicylate. A coloured solution is obtained. The intensity of this colour is measured at 540 nm. The intensity is directly correlated to the concentration.

A set of standard starch solutions is prepared. The intensities of these solutions are measured using a spectrophotometer. A standard graph is plotted taking concentration on the X-axis and intensity on the Y-axis. From the intensity of the test sugar solution, the concentration is determined.

$$\% \text{ Starch} = \frac{\text{Amount of glucose}}{\text{Amount of starch}} \times 0.90 \times 100$$

Polarimeter method   A polarimeter is an instrument used to measure the optical rotation of an optically active substance. Sugars are optically active substances. The extent of optical rotation is proportional to the concentration. This principle is applied in this method.

Note   Similar to this polarimeter method; other methods like refractometer and densitometer methods are applied to quantitative determination of the sugars.

REVIEW QUESTIONS

Give short answers

1.   Define carbohydrate.

2.   Draw the structure of aldotriose.

3.   What is the structure of a ketotetrose?

4.   Dextrin is a type of _____.

5.   Lactose is composed of _____ and _____.

6.   Name the two anomers of glucose.

7.   Draw the ring structure of fructose.

8.   Arrange the following sugars in the increasing order of sweetness:

   Galactose, lactose, fructose, glucose, sucrose.

9. What is fermentation of sucrose?

10. Complete the following reaction:

$$?? \xleftarrow{\text{H}_2\text{O, total}} \text{starch} \xrightarrow{\text{H}_2\text{O, partial}} ??$$

11. Write any two uses of pectin.

12. What are the two long-chain polymers that make starch?

13. Briefly explain the Molisch Test.

14. What is Bials' test for carbohydrates?

15. Write the chemical equation of Benedict's test.

Give detailed answers

1. Explain the classification of carbohydrates with two examples each; outline their nutritional sources.

2. How does glucose react with each of the following?

   i. AC_2O

   ii. H_2, Ni

   iii. $PhNH-NH_2$

   iv. HNO_3

   v. Br_2, H_2O

3. Explain the reaction of sucrose with the following:

   i.   Water, acid

   ii.   Sulphuric acid

   iii.   Yeast

   iv.   HCl

   v.   Nitric acid

4. Draw the ring structures of the following:

    i.   β-D-Glucopyranose

    ii. α-D-Glucopyranosyl-β-D-Fructofuranose.

5. Write the principle of manufacture of sucrose.

6. Arrive at the structure of starch by chemical methods.

7. Discuss the biological functions of carbohydrates.

8. Write the summary of digestion of carbohydrates.

9. Explain the metabolism of carbohydrates.

10. How are sugars biosynthesized?

11. Explain the chemistry of the following qualitative tests of sugars:

    i.   Benedict's test

    ii. Seliwanoff's test

    iii. Iodine test

    iv. Fehling's test

12. Discuss the reduction method of estimation of sugar.

13. How is starch estimated quantitatively?

PROBLEMS

1. A food sample (3.3 g) is treated with 50 mL ethanol in order to dissolve the sugar present in it. It is made up to 100 mL. Exactly 10 mL of the made up solution is treated with a required amount of copper sulphate to give copper (I) oxide. This red solid mass is dissolved in ferric alum and the resulting solution is titrated with a known quantity of potassium permanganate. The amount of reducing sugar is 0.04172 g. Calculate the percentage of reducing sugar.

Ans: 12.64

2. 2.9 g of a food sample is treated with 10 mL dilute HCl. Then, the sugar is extracted with ethanol and made up to 200 mL. 10 mL of the made up solution is treated with a sufficient amount of copper sulphate. The copper (I) oxide formed is dissolved in ferric alum and titrated with a known quantity of potassium permanganate solution. The amount of copper is found to be 0.02412 g. Calculate the percentage of total sugar.

Ans: 16.63

3. A fruit sample (1.080 g) is taken for the analysis of sugar by Bertrand's method. It is dissolved in a solvent to give 200 mL solution. Exactly 20 mL each of two aliquots is taken for estimation of total sugar (0.1510 g) and reducing sugar (0.01125 g). Find out the percentage of non-reducing sugar.

Ans: 6.521

4. A fruit sample (1.305 g) is dissolved in 100 mL water–ethanol (1 : 1) solvent mixture. A standard glucose solution

is prepared by dissolving 5.003 g glucose in 100 mL water in a standard volumetric flask. The standard solutions of different volumes are taken in different test tubes. Each of the standard solutions (0.5, 1.0, 1.5, 2.0, 2.5 and 3.0 mL) is mixed with 2 mL 3, 5-dinitrosalicylate reagent and the total volume is made up to 10 mL. The % transmittance of each of the standard solutions is measured at 540 nm. In a similar manner, 1 mL of the fruit sample solution is mixed with the reagent and the transmittance is measured. The data are given below:

Std. soln.	0.5	1.0	1.5	2.0	2.5	3.0	Test soln.
% T	99	94	88	80	74	71	86

Draw the standard graph. Find out the percentage of sugar present in the fruit sample.

A known quantity of a fruit sample is treated with HCl. After the hydrolysis of the non-reducing sugars, it is neutralized and dissolved in 250 mL water–ethanol solvent mixture. Exactly 20 mL of this solution is taken for sugar analysis by Bertrand's method. The amount of sugar is 74 mg. The percentage of sugar is 14.1894. What is the actual amount of the fruit sample?

Ans: 6.521

3

PROTEINS AND AMINO ACIDS

INTRODUCTION

Proteins are the principal nitrogenous constituents of protoplasm of all tissues of plants and animals. They are necessary for the synthesis of body tissues and for many regulatory functions. Amino acids are the fundamental building units of proteins and they contain carbon, nitrogen, oxygen, phosphorus, sulphur, etc. as the constituent elements. Many amino acids combine through the peptide linkage to form proteins.

Some proteins contain small amount of metal ions such as iron (II), copper (II), calcium (II), magnesium (II), etc. Proteins are biopolymers that are different from carbohydrates and fats in having nitrogen as the main constituent element. The molecular mass of proteins may range from 10^4–10^6. As they form colloids, they do not readily pass through membranes.

Proteins are very specific in the number of amino acids and in the way these acids are sequenced. It is this specificity that gives various tissues their unique nature in the functions and characteristics.

CLASSIFICATION OF PROTEINS

Proteins are classified based on the general structure and the chemical nature. There are three types of proteins—simple proteins, compound proteins and derived proteins.

Simple proteins contain only amino acids and their derivatives. **Compound proteins** have simple proteins along with some non-protein parts. **Derived proteins** are polypeptides in different fragments depending upon the size. The fragments vary from small to large in size. Table 3.1 gives the details of these types of proteins.

Proteins are classified into seven types based on their functions. They are **structural proteins** (collagen), **contractile proteins** (muscle), **antibiotics** (gamma globulin), **blood proteins** (albumin), **hormones, enzymes** and **nutrient proteins** (food sources of essential amino acids).

Table 3.1   Types of proteins and examples

Types	Examples	Nutritional sources
Simple proteins	Albumin Globulin Glutelins Albuminoid	Milk, blood, egg, blood serum, glutens in wheat, collagen, keratin, gelatin
Compound proteins	Nucleoproteins Glycoproteins Phosphoproteins Lipoproteins Metalloproteins	Glandular tissues, mucin (secretions of mucous membrane), casein, phospholipid, cholesterol, haem of haemoglobin
Derived proteins	Fragmented proteins	Proteoses, peptones, polypeptides

AMINO ACIDS

Amino acids are a type of organic compounds having two functionality –COOH and NH_2 groups. Twenty-one amino acids are known. Of these, nine are called essential amino acids and the other twelve non-essential amino acids. The list of amino acids is given in the Table 3.2.

Table 3.2   List of amino acids and their abbreviations

Essential amino acids	R	Non-essential amino acids	R
Threonine (Thr)	$CH(OH)CH_3$	Glycine (Gly)	H
Valine (Val)	$CH(CH_3)_2$	Alanine (Ala)	CH_3
Leucine (Leu)	$CH_2CH(CH_3)_2$	Serine (Ser)	CH_2OH
Isoleucine (Ile)	$CH(CH_3)CH_2CH_3$	Cysteine (Cys)	CH_2SH
Methionine (Met)	$(CH_2)_2SCH_3$	Aspartic acid (Asp)	CH_2COOH
Lysine (Lys)	$(CH_2)_4NH_2$	Glutamic acid (Glu)	$(CH_2)_2 COOH$
Phenylalanine (Phe)	$C_6H_5CH_2$	Hydroxylysine (Hyl)	$(CH_2)_2–CHOH–CH_2– NH_2$
Tryptophan (Trp)		Tyrosine (Tyr)	
Histidine (His)		Proline (Pro)	
Arginine (Arg)	$(CH_2)_3NHC(=NH)NH_2$	Hydroxyproline (Hyp)	
		Asparagine (Apr)	

The structures of some amino acids are given in Figure 3.1. The amino acids are generally represented by the formula $NH_2-CH(R)-COOH$. The R- can be a group like CH_3, $-CH_2OH$, $-CH_2Ph$, etc.

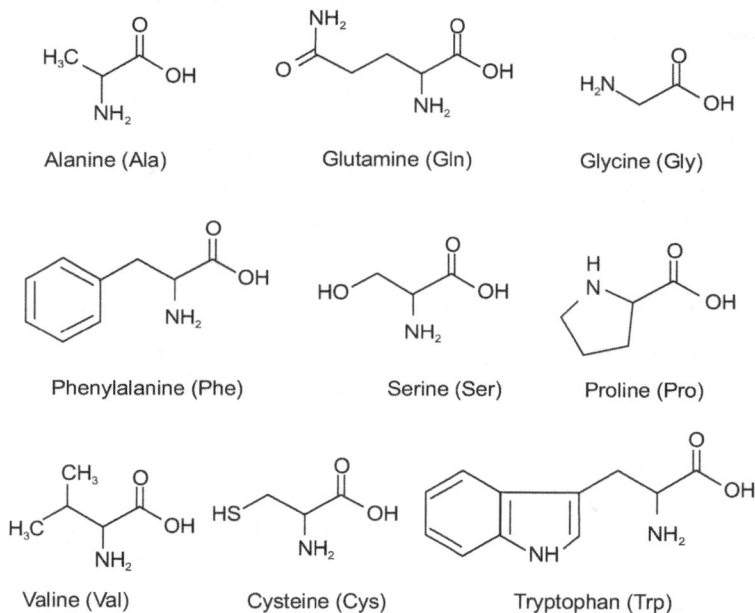

Alanine (Ala)   Glutamine (Gln)   Glycine (Gly)

Phenylalanine (Phe)   Serine (Ser)   Proline (Pro)

Valine (Val)   Cysteine (Cys)   Tryptophan (Trp)

Figure 3.1   Structures of some amino acids

Generally, amino acids are neutral as they have at least one $-NH_2$ and one $-COOH$ group. But, some amino acids are basic and some are acidic. This is because of the presence of one extra $-NH_2$ or one extra $-CO OH$ group. Suppose, the amino acid has one extra amino group, then it is called basic amino acid. If the amino acid contains one extra carboxylic acid group, then it is called acidic amino acid.

Dipolar Nature of Amino Acids

Amino acids exist in the ionic or dipolar form. It is otherwise called zwitterionic form (Figure 3.2). The existence of this ionic form is supported by the following properties of amino acids.

✳ Amino acids are crystalline solids and soluble in highly polar solvents.

✳ Amino acids have high melting points.

✳ They are weakly acidic due to the $-NH_3^+$ group and weakly basic due to the $-COO^-$ group.

✳ They do not show all the chemical reactions characteristic of amino functionality or carboxylic acid functionality.

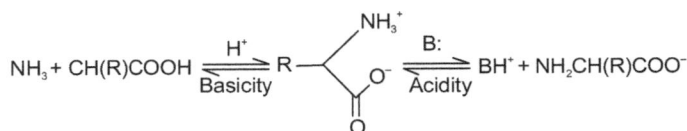

$$NH_3 + CH(R)COOH \underset{Basicity}{\overset{H^+}{\rightleftharpoons}} R-C\begin{smallmatrix}NH_3^+ \\ O^- \\ \| \\ O\end{smallmatrix} \underset{Acidity}{\overset{B:}{\rightleftharpoons}} BH^+ + NH_2CH(R)COO^-$$

Figure 3.2 The zwitterionic form of an amino acid: acidity and basicity

An amino acid can exist in three different ionic forms depending upon the pH of the medium. But, at a particular pH, the amino acid is electrically neutral; that situation is called **isoelectric point**. Considering glycine as an example, we can understand the concept of isoelectric point through the titration curve (Figure 3.3).

$$H_3N^+CH_2COOH \underset{+H^+}{\overset{-H^+}{\rightleftharpoons}} H_3N^+CH_2COO^- \underset{+H^+}{\overset{-H^+}{\rightleftharpoons}} H_2NCH_2COO^-$$

Low pH   Neutral pH   High pH

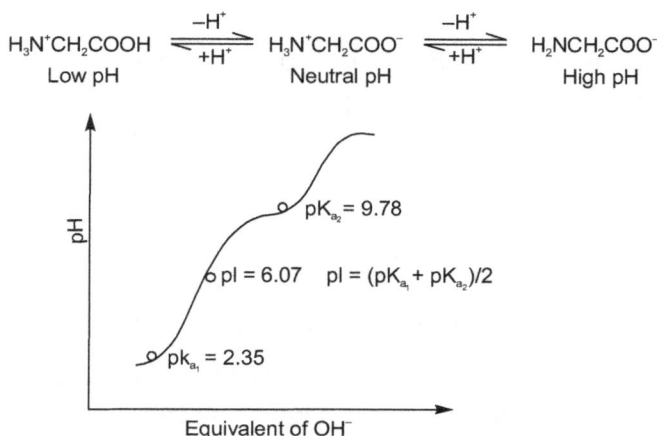

Figure 3.3   Titration curve of glycine

Synthesis of Amino Acids

Amino acids can be synthesized under laboratory conditions by several methods. Three simple methods are discussed here.

Perkin's method   Acetic acid is treated with bromine in the presence of phosphorus to get bromoacetic acid. It is further treated with excess ammonia to form glycine.

$$CH_3COOH \xrightarrow{Br_2,P} BrCH_2COOH \xrightarrow{Excess\ NH_3} H_2NCH_2COOH$$

Strecker's method   α-phenyl acetic acid reacts with sodium cyanide in the presence of ammonium chloride to form amino-cyanide compound. Then, it reacts with dilute acid to yield phenylalanine.

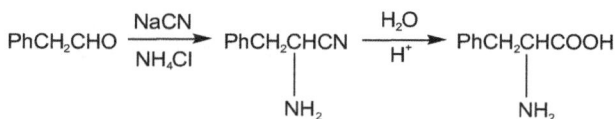

$$PhCH_2CHO \xrightarrow[NH_4Cl]{NaCN} \underset{\underset{NH_2}{|}}{PhCH_2CHCN} \xrightarrow[H^+]{H_2O} \underset{\underset{NH_2}{|}}{PhCH_2CHCOOH}$$

Gabriel's method Malonic ester forms its bromo derivative on reaction with bromine in CCl_4. It reacts with potassium phthalimide and subsequently undergoes acid and base hydrolysis to yield phenylalanine and phthalic acid.

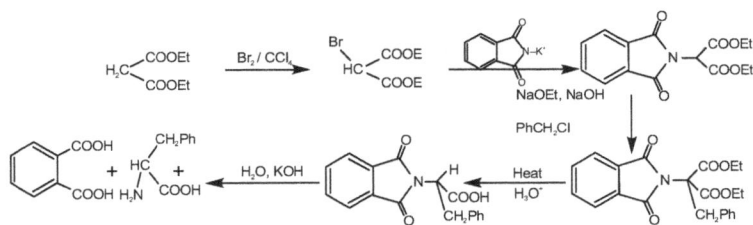

SYNTHESIS OF PROTEINS OR POLYPEPTIDES

Emil Fischer first synthesized peptides. His method of preparation consists of three stages.

1. Protecting the amino group

2. Forming the peptide linkage

3. Removing the protecting group

This sequence of steps is repeated several times depending upon the peptide that is to be prepared.

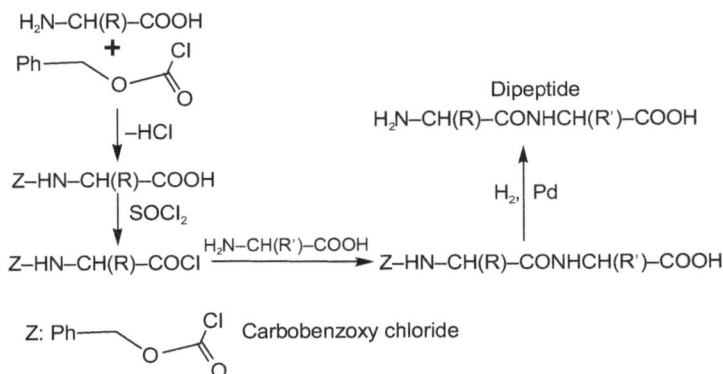

The –CONH– linkage is called **peptide linkage**. A protein molecule has many such peptide linkages. That is why proteins are called polypeptides.

STRUCTURE OF PROTEINS

In a peptide linkage the C–N bond has predominantly double bond character. The carbonyl carbon, nitrogen and the other atoms in a peptide lie in a plane as shown below:

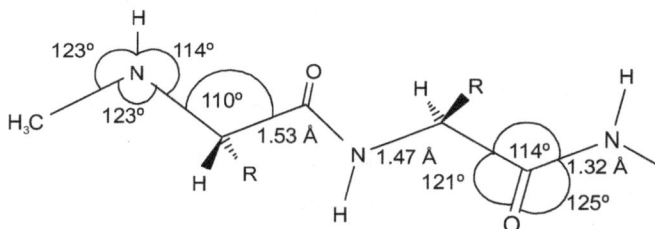

Proteins are found to have three types of structures:

   i.   Primary structure

  ii.   Secondary structure

 iii.   Tertiary structure

The primary structure of a protein is a linear chain structure. This chain is made up of many amino acids sequenced to a polypeptide through peptide linkages.

The secondary structure of protein is a helical structure. The helix may be of alpha- or beta-type. This helical structure is formed by hydrogen bonding between the $-C=O$ functionality of one amino acid residue and –NH functionality of another amino acid residue.

The tertiary structure of the protein is a globular structure. It is also formed by the hydrogen bonding between the $-C=O$ functionality of one amino acid residue and $-NH$ functionality of another amino acid residue in different polymeric peptide chains. These structures are shown in Figure 3.4.

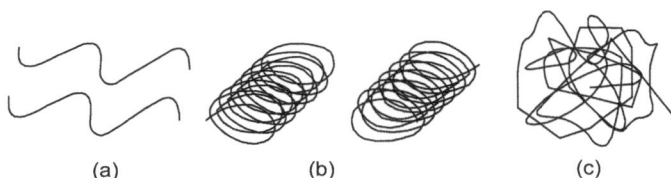

| (a) | (b) | (c) |

Figure 3.4 Structures of proteins   (a) Linear chain [primary structure]   (b) Alpha-and beta-helices [secondary structure]   (c) 3D-Globular structure [tertiary structure]

GENERAL PROPERTIES

Mostly, proteins are colourless and tasteless solids, insoluble in water. But, they are soluble in alcohol and ether. Proteins are complex molecules having several thousands of atoms. Generally they have high molecular mass (Table 3.3).

Hydrolysis

Proteins ⟶ Simpler proteins ⟶ Amino acids

This hydrolysis can be effected by dilute mineral acid, HCl or by the enzyme, protease.

Table 3.3 Molecular masses of some proteins and their sources

Proteins	Mol. mass
Insulin (pancreas)	12000
α-Lactalbumin (milk)	15000
Chymotrypsinogen (bovine)	23500
Zein (corn)	40000
Fibroin (Silk)	56000
Urease (enzyme)	480000
Casein (milk)	22000
Lactoglobulin (milk)	36000
Haemoglobin (blood)	65000
Hexokinase (yeast)	100000
Glutamin (wheat)	110000
Tobacco mosaic (virus protein)	41×10^6

Colloidal nature As proteins are heavy molecules, they form colloids when dispersed in water medium. This gel formation is due to the binding of the hydrophilic ends of proteins with the polar ends of water molecules.

Amphoteric character Proteins can react with both acids and bases.

Isoelectric point Like the amino acids, proteins exhibit electrical neutrality at a particular pH. This point of pH is called the **isoelectric point** (Table 3.4).

Table 3.4 Isoelectric point (pI) of some proteins

Proteins	pI	Proteins	pI
Casein	4.6	Lysozyme	11
Egg albumin	4.7	Cytochrome-*c*	10
Insulin	5.3	Haemoglobin	6.8
Pepsin	1.1 (approx.)	Fibrinogen	5.5

Denaturation Proteins undergo drastic changes in their behaviour when they are subjected to heating or reaction with strong acids, bases or other reagents, and lose their physiological functions. This is termed **denaturation**.

For example, egg turns into a solid on heating.

Optical activity All proteins show optical activity. It is the ability to rotate the plane of polarized light. All proteins have any one of R, S, D or L configurations.

TERMINAL RESIDUE ANALYSIS

Proteins are analysed to know the exact sequencing of amino acids. This analysis is done by chipping off one amino acid residue from one end of the polypeptide chain. There are two main approaches: N-terminal analysis and C-terminal analysis.

Sanger's method It is an N-terminal analysis. The reagent used is 2,4-dinitrofluorobenzene, DNFB.

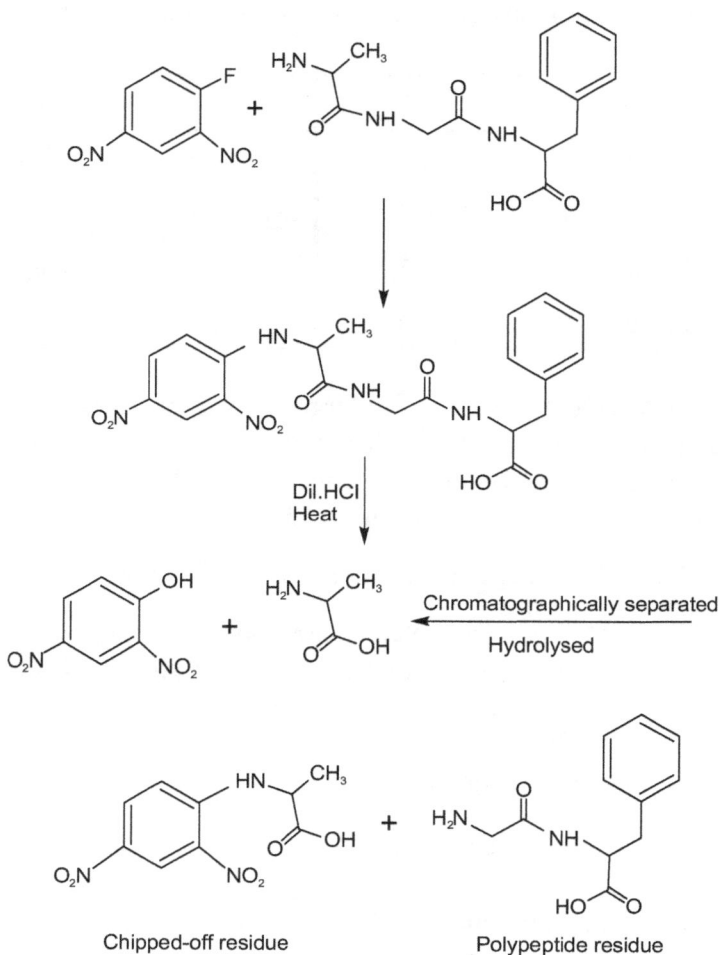

Chipped-off residue   Polypeptide residue

Pher Edman's method It is also an N-terminal analysis. The reagent used is phenyl isothiocyanate.

Reduction method I It is a C-terminal analysis. The reagent used is lithium borohydride (LiBH$_4$). In this method, the C-terminal amino acid residue alone is chipped off as amino alcohol. The rest of the polypeptide is left intact.

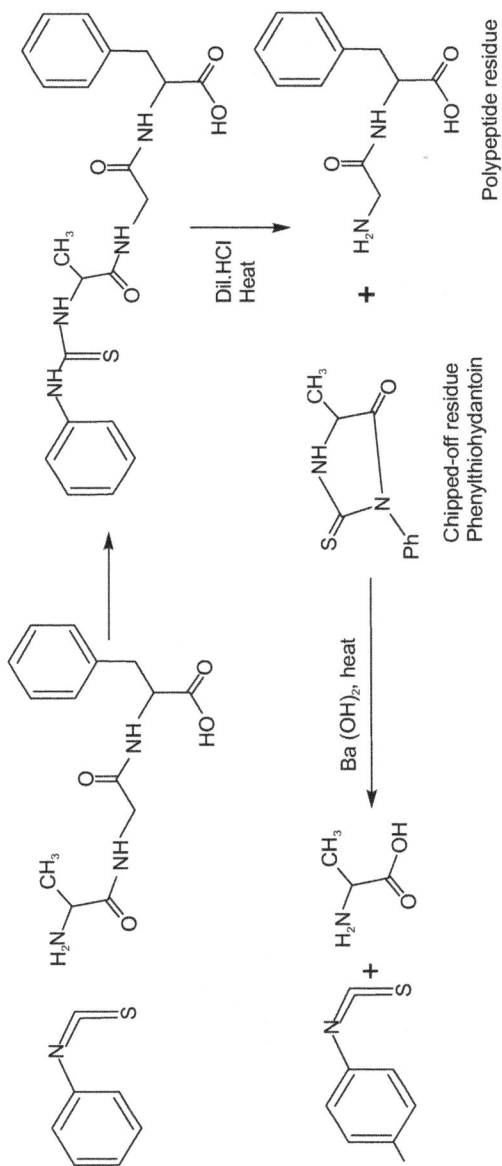

Dil.HCl
Heat

Polypeptide residue

+

Chipped-off residue
Phenylthiohydantoin

Ba (OH)$_2$, heat

+

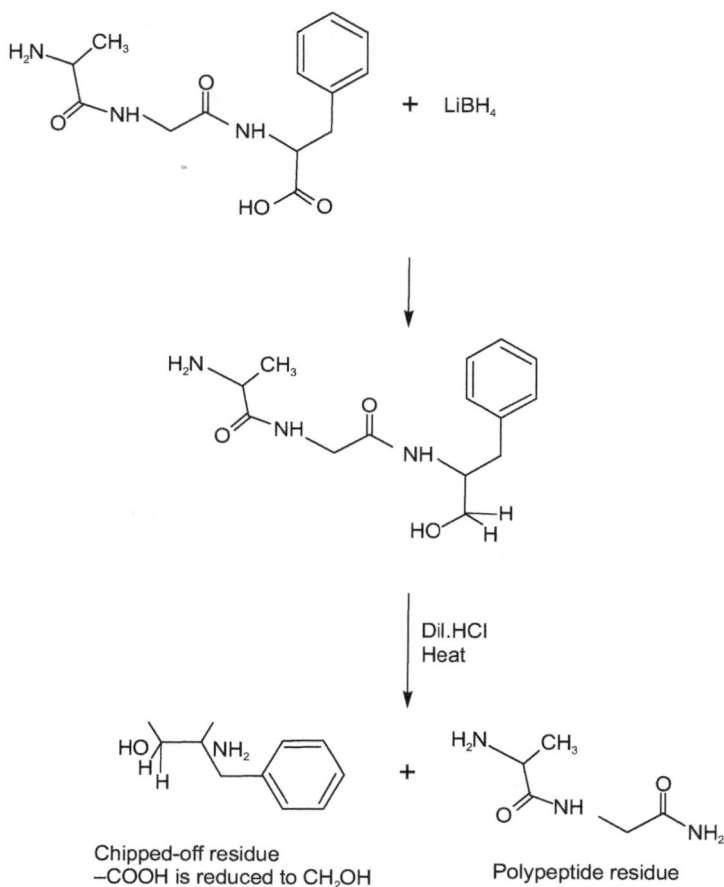

Chipped-off residue
—COOH is reduced to CH₂OH

Polypeptide residue

Reduction method II   It is also a C-terminal analysis. The reagent used is hydrazine (H_2N–NH_2). Here, the C-terminal amino acid residue alone is chipped off. The rest of the residues are cleaved off in the reduced form.

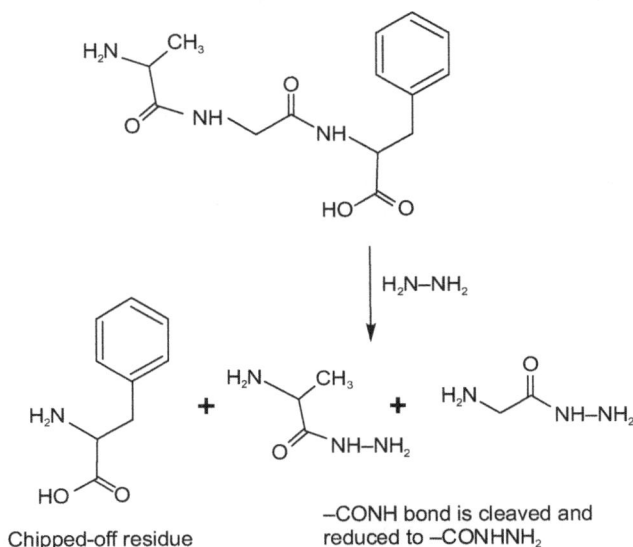

Chipped-off residue

−CONH bond is cleaved and
reduced to −CONHNH$_2$

Enzyme method Here, the enzyme used is carboxypeptidase. It cleaves the C-terminal residue alone and leaves the rest of the chain undisturbed. Gradually, one by one, the C-terminal residues are chipped off.

FUNCTIONS OF PROTEINS IN THE BODY

The primary function of proteins in living cells is to promote growth and maintenance of tissues. From the amino acids, the specific cellular tissue proteins (enzymes and hormones) are synthesized.

Physiological and Metabolic Roles

1. *Methionine* is a methylating agent. It participates in the formation of some non-protein cellular constituents like choline.

2.   *Tryptophan* acts as a precursor of niacin and serotonin.

3.   *Phenylalanine* is a precursor of the hormone thyroxine and epinephrine.

Contribution to the energy metabolism   When the nitrogenous part of the amino acids is removed, the rest of the residue may be either glycogenic (able to form carbohydrates) or ketogenic (able to form fatty acids). Leucine, phenylalanine and tyrosine are completely ketogenic. The other amino acids are glycogenic. Thus, the carbohydrates formed are oxidized to yield energy.

DIGESTION OF PROTEINS

Proteins undergo mechanical digestion in the mouth. It is turned into a semi-solid and mixed with the saliva.

Chemical digestion begins only in stomach. Gastric HCl activates the pepsinogen produced in the stomach, thereby **pepsin** is formed. This active pepsin breaks the peptide linkages of proteins. Thus, short-chain polypeptides, **proteoses** and **peptones** are produced, first. There is another gastric enzyme, **rennin**, it helps the digestion of casein in milk. It is important in an infant's digestive process. Rennin is absent in adult's digestive process.

A number of pancreatic and intestinal enzymes are involved in the digestion of proteins in the small intestine. The big protein molecules are broken down to simpler peptide chains and in turn to amino acids. The following processes occur in alkaline medium (as contrary to acidic medium in the stomach):

* ※   *Enterokinase* activates trypsinogen to form trypsin.

* ※   *Trypsin* activates chymotrypsinogen to form chymotrypsin.

❊   *Carboxypeptidase,* an enzyme of the peptidase group acts on proteins to form free –COOH groups and amino acids.

❊   *Aminopeptidase,* acts on proteins to produce free –NH$_2$ groups and the amino acids.

❊   *Dipeptidase* acts on all dipeptides to produce free amino acids.

The mechanism of digestion is given in Table 3.5.

ABSORPTION OF PROTEINS

The end products of protein digestion are water-soluble amino acids. These amino acids are absorbed rapidly from the small intestine into the portal blood system. A fine network of villous capillaries does this job of absorption. The absorption mechanism is not known exactly. The olden concept assumed absorption by passive diffusion. But, the modern concept assumes absorption by diffusion, pinocytosis, and active transport. For example,

❊   whole proteins (few fragments) are absorbed as such.

❊   vitamin B$_6$ is absorbed along with amino acids by the active transport mechanism.

METABOLISM OF PROTEIN

The metabolism of protein involves a collection of very complex and closely interconnected chemical activities. These metabolic activities are of three broad categories—balance, building tissues (anabolism) and breakdown of tissues (catabolism).

Table 3.5   Summary of protein digestion

Organ	Inactive form	Activator	Active enzyme	Digestive action
Mouth			None	Mechanical digestion
Stomach	Pepsinogen	HCl	Pepsin	Protein → Simpler peptide chains
			Rennin	Casein → Coagulated curd
Intestine	Trypsinogen	Enterokinase	Trypsin	Protein → Simpler peptide chains, amino acids
Pancreatic juice	Chymotrypsinogen	Trypsin	Chymotrypsin	--do--
			Carboxypeptidase	--do--
Intestinal juice			Aminopeptidase	--do--
			Dipeptidase	Dipeptides → Amino acids

Balance

In living things, there exists a balance. There is a constant inflow and outflow of materials, constancy in building up and breaking down of parts. A balanced depositing and mobilizing of constituents occurs. This balanced state is termed as equilibrium or **homoeostasis**. The activities connected to the maintenance of this equilibrium are called **homoeostatic mechanisms**. The balance between the body parts and functions is life-sustaining. In the growth process, the rate of building up is more than the rate of breaking down. In case of starvation or disease, the destruction is more than construction.

The growth process is identified with the positive nitrogen balance. That is, the intake of protein is higher than the output of protein. It is the reverse for the negative nitrogen balance.

Building of Tissues (Anabolism)

The tissue proteins are synthesized with specificity. The type, number and the sequencing of amino acids are very important in the synthesis of proteins and are controlled by DNA, mRNA and tRNA. control these. The whole process is diagrammatically illustrated in Figure 3.5.

Breaking Down of Tissues (Catabolism)

If an amino acid is not used in protein synthesis, then, it is degraded or oxidized to yield energy. In the catabolism, the amino acids give a nitrogenous and a non-nitrogenous residue. Figure 3.5 illustrates the catabolism of proteins.

Certain hormones—pituitary growth hormone, androgens, insulin and thyroid hormone—regulate anabolism and catabolism of protein.

(a)

DNA—genetic code

Messenger RNA

DNA transfers specific protein
pattern to messenger-RNA

(b)

Cell ribosomes

Activated amino acids in cell
cytoplasm attach to transfer-RNA

(d)

The new polypeptide breaks free;
transfer-RNA is released to repeat
the process

(c)

Transfer-RNA carries the active
amino acids to appropriate
positions and forms peptide links

Carbohydrate and fat

Amino acid$_1$ ⟶ NH$_3$ $\xrightarrow{\text{+Keto acid}}$ Amino acid$_2$

$\xrightarrow{\text{+ Amino acid}_3}$ Amine form of amino acid$_3$

Keto acid
non-nitrogenous
residue

⟶ Purines ⟶ Uric acid

⟶ Urea

Protein   Carbohydrate   Fat   Protein
(glucogenic)   (ketogenic)
                        EMP

Pyruvate ⟶ Active acetate

Oxaloacetate

Krebs' cycle

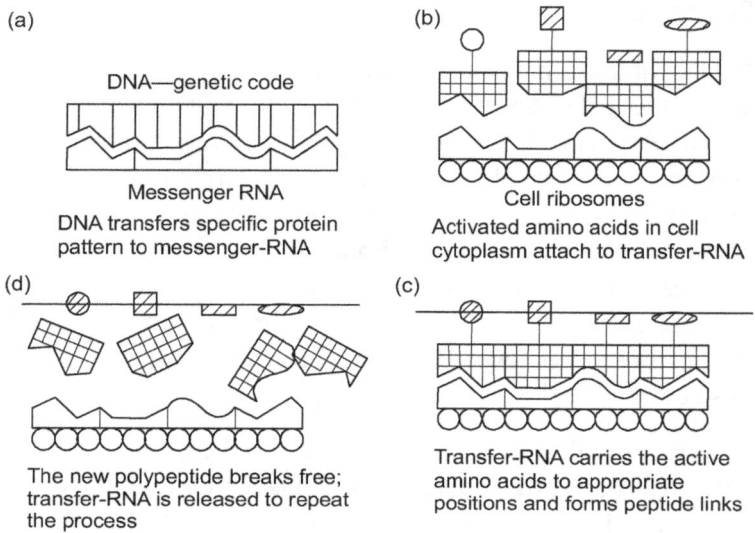

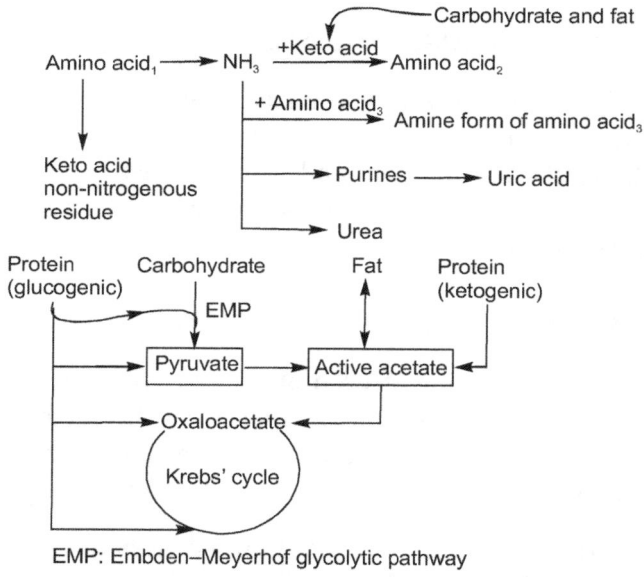

EMP: Embden–Meyerhof glycolytic pathway

Figure 3.5   Mechanism of protein synthesis (anabolism) and
breakdown (catabolism)

PROTEIN REQUIREMENTS AND PROTEIN COUNTER

Some information about the daily protein requirements and the protein-rich sources is given in Table 3.6. These data are mainly applicable for Indian children. The daily requirements of protein may be in the range 22–63 g for the children in the age group of 1–18.

Table 3.6 Approximate daily protein requirements and protein counter

Meal/Food	Protein(g)	Food (Quantity)	Protein (g)
Breakfast		Meat (palm-sized) (28 g)	7
1 small cup of milk	6.4	Fish (palm-sized) (28 g)	7
Lunch		Chicken (palm-sized)	7
1 chappati	3	1 Egg (large size) (50 g)	6
1 small cup of lentils	6	1 Milk cup (250 mL)	8
1 small cup of vegetables	2	Yogurt -do-	8
1 banana	0	Cheese (1 slice) (20 g)	6
Tea		Paneer (50 g)	6
1 small cup of milk	6.4	Lentils (1 small cup)	6
Dinner		Beans (1small cup)	6
2 small cups of rice	6	Nuts (1 small handful)	6
1/3rd palm-sized meat piece	7	Rice (1 small cup)	3
1 small cup of vegetables	2	Vegetables cooked (1 small cup)	2
1 orange	0		
Total Protein	38.8		

(*Source:* Recommended Dietary Allowances for Indians. *Nutrition for the Mother and Child*, 5th Edition. NIN, ICMR, Hyderabad. 2002.)

ANALYSIS OF PROTEINS

Proteins are analysed both qualitatively and quantitatively. There are a number of colour reactions, which are advantageously used for the analyses. The groups like $-NH_2$, $-CONH-$ and $-COOH$ take part in the colour reactions. Proteins are to be free from the chloride and ammonium salt forms, before being used for analysis. Ion-exchange or dialysis does the desalting. The desalted solution is lyophilized and stored at low temperature ($< 0°C$). For example, egg albumin, lactalbumin, lactoglobulin, etc. are desalted and lyophilized.

Qualitative Analysis

The proteins present in any foodstuff can be tested qualitatively by the following tests. The procedures of these tests are discussed briefly.

Biuret test   2 mL protein solution + 2 mL sodium hydroxide solution (10%) + 2 mL copper (II) sulphate solution. Mix well. Warm and cool. Formation of violet colouration confirms protein.

Millon's test   2 mL protein solution + 4 mL of Millon's reagent. Heat and cool. Red colour formation indicates the presence of protein (This is due to the presence of).

Millon's reagent   Mercury dissolved in just sufficient amount of HNO_3. It is diluted with water.

Ninhydrin test   2 mL protein solution + 4 mL ninhydrin. Heat to 100°C in water bath. Blue or purple colouration signifies the presence of protein.

The chemistry of these tests is given below:

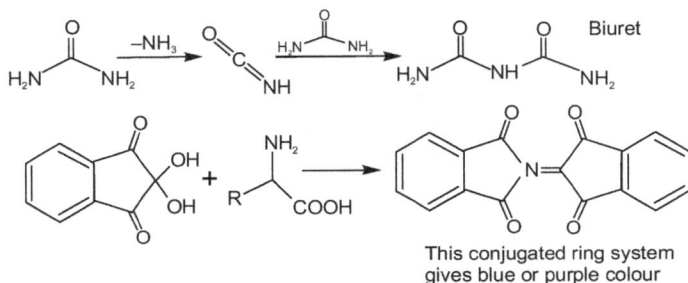

This conjugated ring system gives blue or purple colour

Quantitative Analysis

These analyses help us estimate the amount of proteins present in a foodstuff. The principles of these methods are discussed here.

Biuret method   A standard protein solution is prepared first. Using this bulk standard, protein solutions of five or six different concentrations are prepared. Each of the standard solution is mixed with a definite quantity of biuret reagent. The intensity of these coloured solutions are measured at 550 nm, using a photocolorimeter. A standard graph is plotted taking concentration on the X-axis and the absorption on the Y-axis. The test protein solution is also mixed with the biuret reagent and its intensity is measured. This intensity value is fitted in the standard graph as shown in the figure that follows.

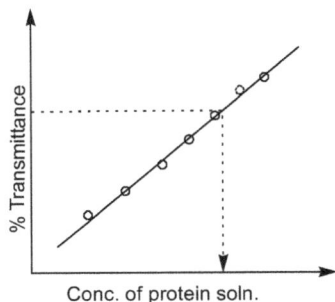

Conc. of protein soln.

Ninhydrin method The protein is dissolved in sulphuric acid. The resulting solution is neutralized with barium hydroxide and its pH is adjusted to 5. The resultant solution is then suitably diluted and the bulk standard solution is prepared. From the bulk, an aliquot of different amounts are taken. Each of these aliquots is mixed with definite amount of ninhydrin. The coloured solutions are used for the photocolorimetric measurements at 570 nm. A standard graph is plotted. The test protein solution is also subjected to this measurement. From the intensity of the test solution, the concentration is obtained.

Kjeldahl's method A known amount of the sample (W_p) is kjeldahlized by heating with HgO, Hg, K_2SO_4 and conc. H_2SO_4. The kjeldahlized sample is treated with NaOH. The liberated ammonia is trapped in boric acid. Then, it is titrated with a standard HCl (N_A) using methyl orange indicator. A blank titration is also done in a similar way without the protein sample. From the titration values for the sample solution (T_s) and blank (T_b), the percentage of nitrogen is calculated using the formula as shown below:

$$\% \text{ Nitrogen} = \frac{(T_b - T_s)N_A \times 14 \times 100}{W_P}$$
$$\% \text{ Protein} = \% \text{ Nitrogen} \times (6.25)$$

REVIEW QUESTIONS

Give short answers

1. What are proteins?

2. How are proteins classified based on their functions?

3. Draw the structures of

   i.   Cysteine

   ii. Phenylalanine

4. What are essential amino acids?

5. What is the zwitterionic form of an amino acid?

6. Complete the following:

   $C_6H_5CH_2CHO \longrightarrow ? \longrightarrow ?$

7. Arrange the following proteins in the decreasing order of molecular mass.

   Haemoglobin, casein, hexokinase, glutamine, tobacco mosaic.

8. What is isoelectric point?

9. What is denaturation of proteins?

10. Are all proteins optically active?

11. Pher Edman's method is employed for _____ analysis.

12. The enzyme used for the C-terminal analysis is _____.

13. Homoeostasis means _____.

14. Outline the following tests:

   i.   Biuret Test

   ii. Millon's Test

Give detailed answers

1. What are the three types of proteins? Give two examples for each, and their nutritional sources.

2. Identify the following as essential and non-essential amino acids and draw their structures: Thr, Lys, Val, Gly, Asp.

3. List the properties of amino acids that support the zwitterionic form.

4. How is phenylalanine synthesized by Gabriel's method? Write the chemistry of synthesis of polypeptide, Gly-Phe-Ala.

5. Briefly discuss the primary, secondary and tertiary structures of proteins. Write the chemistry of Sanger's method of N-terminal analysis.

6. How does hydrazine work in the C-terminal analysis of a polypeptide? Write the chemistry of this analysis.

7. Enumerate the biological functions of proteins.

8. Explain the summary of digestion of proteins.

9. Write the mechanism of protein synthesis.

10. How are proteins catabolized?

11. Discuss the Biuret method of estimation of protein.

12. Explain the Kjeldahl's method of estimation of proteins.

13. What is the principle involved in the ninhydrin method of quantitative determination of protein.

PROBLEMS

1. Accurately 0.100 g of a food sample is subjected to kjeldahlization. The kjeldhalized sample is treated with excess of 40% NaOH. The liberated ammonia is absorbed in 2% boric acid. Then, it is titrated with N/28 HCl using methyl orange indicator. A blank titration is also carried out. Thus, the amount of nitrogen is found to be 6.92×10^{-4} g. Find out the amount of protein and its percentage. [The amount of nitrogen in any protein molecule is about 16%].

    Ans: 0.0433 g, 43.25

2. Exactly 0.675 g of a protein sample is effected kjeldahlization. 5 mL of the kjeldahlized solution (25 mL) is treated with excess of 40% NaOH. The liberated ammonia is trapped in 2% boric acid. Then, it is titrated with 0.1084N HCl using methyl orange indicator. The titration value is 2.8 mL. A blank titration is also carried out to get the titration value of 0.7 mL. Find out the percentage of protein.

    Ans: 14.752

3. A known amount of a fruit sample is taken for protein analysis by Kjeldahl's method. The amount of nitrogen is 6.316×10^{-3} g. The percentage of protein is 4.928. Find out the amount of the fruit sample.

    Ans: 0.801 g

4. A food sample (0.423 g) is dissolved in 100 mL of a suitable solvent. 1 mL of this solution is taken for the

quantitative estimation of protein by Biuret method. It is 1.202×10^{-3} g. Calculate the percentage of protein.

Ans: 27.824

5. A sample of food (0.5013 g) is treated with trichloroacetic acid. The residue obtained is dissolved in 0.1N NaOH and made up to 100 mL in a standard flask. A standard protein solution is prepared by dissolving 0.511 g of albumin in 0.1 N NaOH and made up to 100 mL. Different volumes (0.5, 1.0, 1.5, 2.0, 2.5 and 3.0 mL) of the protein solutions are pipetted out into different test tubes. To each of the protein solution, 2 mL of biuret reagent is added and the total volume is made up to 10 mL. The % transmittance of each solution is measured at 540 nm. The food sample solution (1.0 mL) is also taken for this measurement. The data are given below. Draw the standard graph and find out the percentage of protein.

Std. soln	0.5	1.0	1.5	2.0	2.5	3.0	Test soln.
% T	85	83	81	79	77	75	80

4

LIPIDS

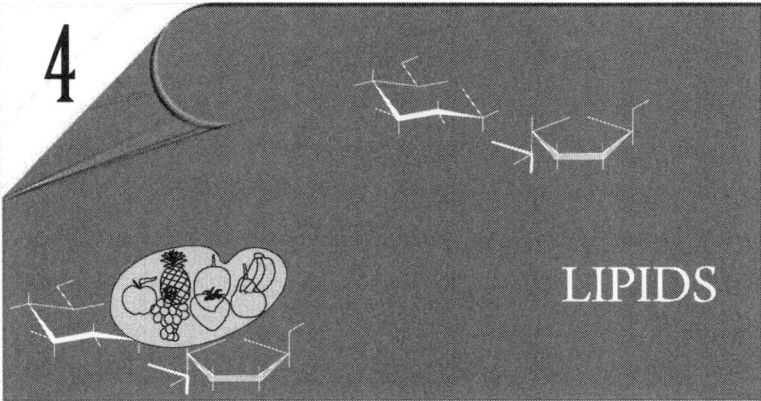

INTRODUCTION

Lipids are a concentrated source of energy for living cells. Lipids supply two-fifths of the total calorie intake. The name "lipid" is derived from the Greek word *lipos,* which means fat. Fats have diverse functions—insulation, padding, cell-membrane integrity, transport of fat-soluble vitamins, hormone synthesis, etc. Lipids are a group of organic compounds—fats, oils, waxes and other related compounds—that are greasy to touch and insoluble in water. They have C, H, O, N, P, etc. as the constituent elements. Lipids are chemically defined as the esters of fatty acids and alcohols. The fatty acids are the long-chain aliphatic part with carboxylic acid functionality.

CLASSIFICATION OF LIPIDS

Lipids are broadly classified into three types based on the ideas of Bloor and Deuel. They are:

1. Simple lipids
2. Compound lipids
3. Derived lipids

Table 4.1 Types of lipids and their chemical nature

Type of lipid	Chemical nature	Examples	Sources
Simple lipids			
Oils and fats	Fatty acid + alcohol	Triglycerides	Vegetable oils, animal fat, meat, fish, etc.
Compound lipids			
Phospholipids	Fatty acid + alcohol + phosphate group	Lecithin, cephalin, lipositol, etc.	Living cells
Glycolipids	Fatty acid + alcohol + carbohydrate	Cerebroside	Living cells
Lipoproteins	Lipids with proteins	Natural fats, phospholipids, etc.	Living cells
Waxes	Fatty acid + aliphatic alcohol		No nutritional source
Derived lipids			
Sterols	Cyclophenanthrene + secondary alcohol	Cholesterol, ergosterol, etc.	Plants and animals

Table 4.1 gives the classification, chemical nature and sources of lipids. Simple lipids are esters of alcohol and fatty acids. Compound lipids are also esters of alcohol and fatty acids with one more organic group. Derived lipids are substances obtained from simple and compound lipids.

FATTY ACIDS

Fatty acids are long-chain carboxylic acids. The carbon chain may be $C_{10} - C_{24}$. There may be unsaturation points, one or more in number in between the long aliphatic chains. They are called **unsaturated fatty acids**. The fatty acids with no unsaturation are called **saturated fatty acids**. The fatty acids are generally represented as R–COOH. The group R is the aliphatic long chain. Some of the aliphatic chains and their trivial names are listed below.

C_n	Name	C_n	Name
C_{10}	Capric	C_{18} (1 unsat.)	Oleic
C_{12}	Lauric	C_{18} (2 unsat.)	Linoleic
C_{14}	Myristic	C_{18} (3 unsat.)	Linolenic
C_{16}	Palmitic	C_{20} (3 unsat.)	Arachidinic
C_{18}	Stearic	C_{20} (4 unsat.)	Arachidonic
C_{20}	Arachidic		

Mostly, the lipids from animal sources are saturated and the lipids from plant sources are unsaturated (Figure 4.1). Some fatty acids are called essential and the others are non-essential. The essential fatty acids are not synthesized by the body. They are obtained through diet. Absence of such essential fatty acids may cause a specific deficiency disease. For example, a type of eczema in infants is due to the deficiency of the fatty acid, linoleic acid. Other fatty acids that the body can synthesize are termed non-essential fatty acids.

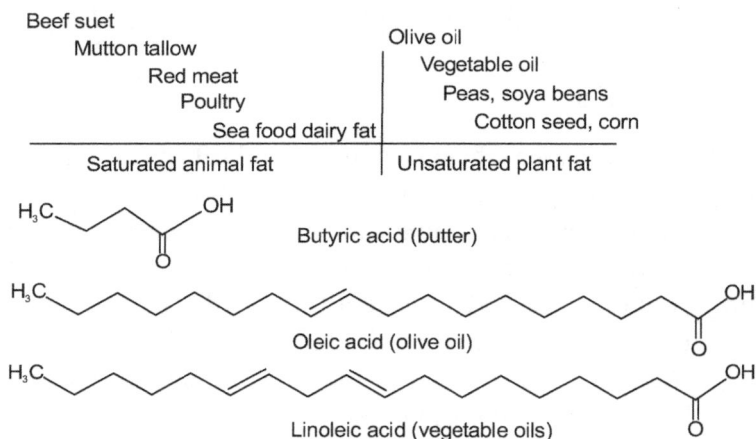

Figure 4.1 Spectrum of food fat according to the degree of saturation of fatty acid constituent and some fatty acids

GENERAL PROPERTIES

Lipids are generally colourless, odourless and tasteless. But, colour, odour and taste are imparted by some other chemical substances that are present along with lipids. For example, the butter is yellow due to its carotene content. Oils are liquids, whereas fats are solids. They are less polar. So they are soluble in less polar solvents. They form emulsion when agitated with water in the presence of soap or gelatin (emulsifier). They are poor conductors of heat and electricity. So, they act as very good insulators for animals. Let us look into some of the chemical properties of simple lipids.

The triglyceride undergoes hydrolysis to give glycerol and fatty acids. The triglyceride gives soap (sodium salt of the fatty acid) and glycerol on reaction with sodium hydroxide. Copper chromite reacts with triglyceride to give glycerol and primary alcohol. Reduction of triglyceride by hydrogen and nickel (catalyst) yields hydrogenated triglyceride (the unsaturated positions of the fatty acid chain get hydrogenated). When triglyceride reacts with oxygen in the presence of sunlight, we get the unsaturated position of the fatty acid chain saturated with peroxy linkages.

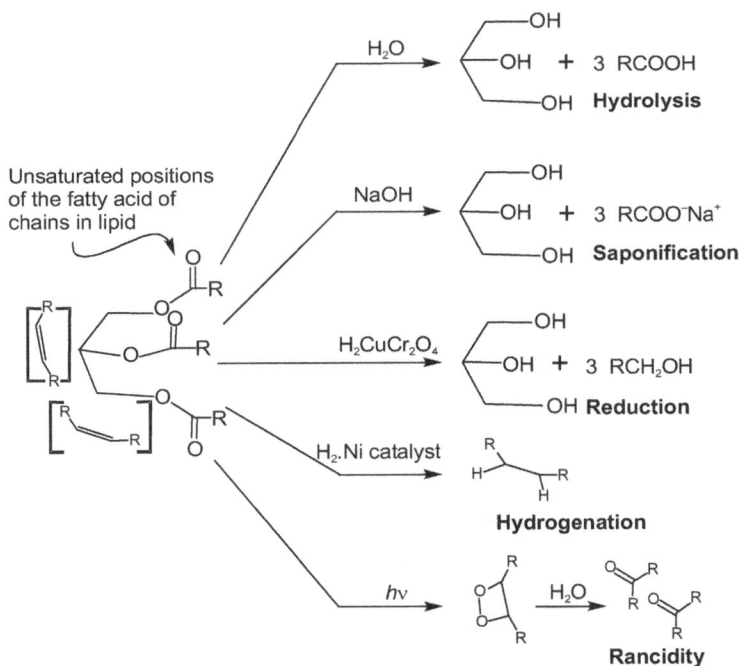

Note Hydrogenation is commercially a useful process. The liquid oils are converted into solid fats by this process. This helps easy transportation.

Saponification is a process by which soap is manufactured. Soaps are different from detergents (sodium salts of sulphonic acids with long-chain substituent or salts of N-substituted anilines having long-chain substituents). Oils acquire foul smell on long exposure to atmospheric air and moisture. This phenomenon is called rancidity.

Structures of Some Lipids

Simple lipids They are glyceride esters, and are derivatives of glycerol and fatty acids.

Compound lipids Phosphatidic acid is the parent structure of phospholipid. Phospholipids contain glycerol, fatty acid and the phosphate group. Sphingolipids are also a type of phospholipids. The structures are shown in the Figure 4.2.

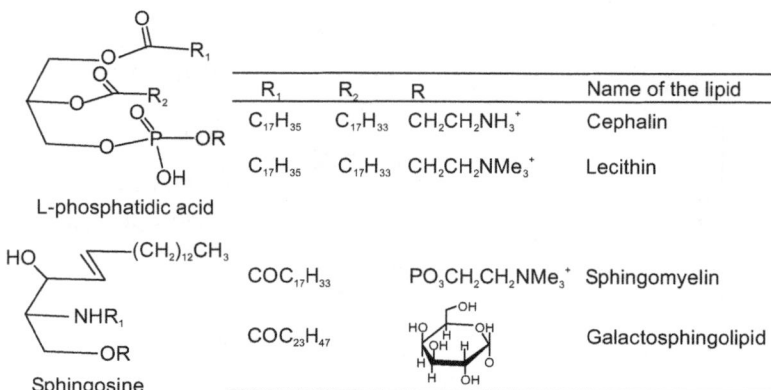

R_1	R_2	R	Name of the lipid
$C_{17}H_{35}$	$C_{17}H_{33}$	$CH_2CH_2NH_3^+$	Cephalin
$C_{17}H_{35}$	$C_{17}H_{33}$	$CH_2CH_2NMe_3^+$	Lecithin
$COC_{17}H_{33}$		$PO_3CH_2CH_2NMe_3^+$	Sphingomyelin
$COC_{23}H_{47}$			Galactosphingolipid

Figure 4.2   Structures of some derived lipids

Lecithins and cephalins have the functions of biological surfactants. They have hydrophilic polar end and hydrophobic non-polar part. These compounds are the constituents of cell walls and membranes.

Derived lipids Steroids are biologically important natural products widely distributed in animals and plants. Some of them are sterols, bile acids, sex hormones, cardiac aglycones, etc.

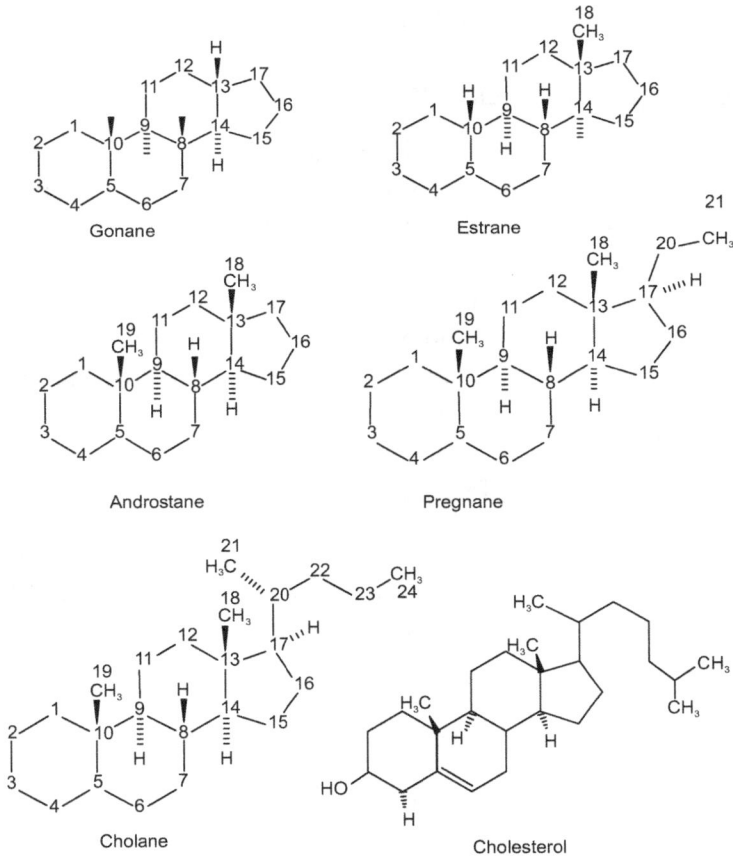

Figure 4.3 Structures of some basic steroids

Steroid hormones are highly specific in their biological functions. Cholesterol is synthesized by most of the animals. This compound is the precursor for the synthesis of other important steroids. Humans often produce more cholesterol than is needed. Then the excess is excreted or collected in the body as gallstones or deposited in the arteries (atherosclerosis). Bile is a mixture of cholesterol and derived steroid amides. Some of the steroid compounds are given in Figure 4.3.

Hydrogenation of Oils

Vegetable oils are usually liquids as they contain more unsaturation ($C=C$ positions). When the oils are hydrogenated, the unsaturation positions are removed and they get saturated, thereby transforming the liquid oils into solids. This is done for certain commercial reasons—for increasing the shelf life, packagability, storage and transportability. But, hydrogenation makes the oil less digestible. The process of hydrogenation is highlighted in Figure 4.4.

FUNCTIONS OF LIPIDS IN THE BODY

Fats serve two basic functions in the human body. One is primary metabolic function to produce energy. Second is the secondary mechanical or structural function to protect a vital organ.

Metabolic Functions

The primary function of lipids is to produce energy. Fats can produce energy twice that produced by carbohydrates or proteins. Fats are highly concentrated and dense. They are weakly soluble. So, they are stored as adipose tissue. The fats in the adipose tissues are used whenever there is a need.

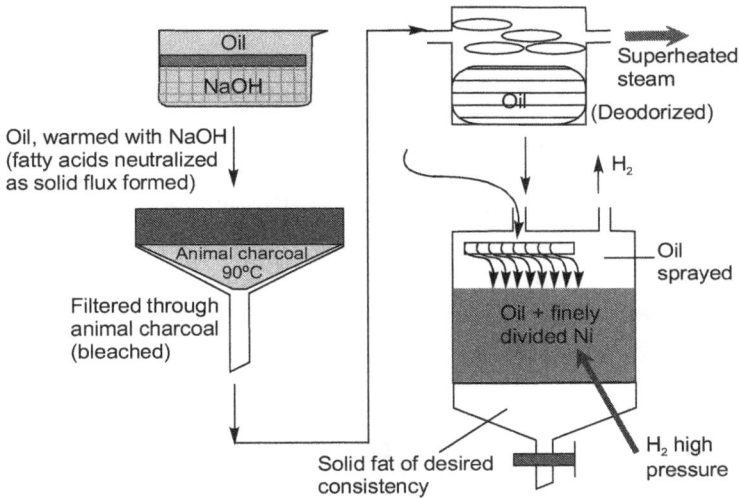

Figure 4.4   Hydrogenation of oil

Mechanical or Structural Functions

Fat provides a general padding for vital organs and for nerves that hold them in place, and helps to absorb shocks. The subcutaneous layer of fat protects the entire body. It insulates the body against rapid temperature changes or excessive heat loss. The main functions of lipids are the following.

Passage of neutral fats   Phospholipids help the neutral fats pass through the cell, the absorption of fat from the intestine and form the essential constituents of nerve tissues.

Sex hormones   Cholesterol and other fat-related compounds help the synthesis of sex hormones and other adrenal hormones.

Palatability and satiety   Fat improves palatability of the food by adding flavour and taste to it. It gives satiety value to the food.

DIGESTION OF LIPIDS

The fat is digested mechanically in the mouth. It is crushed into fine particles, mixed with the saliva and passed on to the stomach. In the stomach the crushed fat is mixed with the secretions. No significant amount of any specific enzyme to digest fat is found in the stomach. But enzymes like gastric lipase and tributyrinase act on emulsified butterfat. It is only in the small intestine that the fat is digested chemically. Secretions from liver, gall bladder, pancreas and small intestine help breaking down the fat.

Bile juices   The bile juice from the liver and gall bladder emulsifies the fat. At this stage, the fat is sized down to small globules and at the same time the surface tension is lowered. The medium is made alkaline.

Pancreatic enzyme   The enzyme lipase (steapsin) acts upon the fat and breaks off the fatty acids one by one from the glyceride esters. Then, cholesterol is combined with fatty acids to form cholesterol esters by cholesterol esterase. These cholesterol esters are in an easy absorbable form.

Intestinal enzyme   The enzyme lecithinase acts upon phospholipids and breaks them into glycerol, fatty acids, phosphoric acid and choline.

Some remaining fats may be pushed to the large intestine and eliminated as faecal fat. The complete summary of digestion of lipids is given in Table 4.2.

Table 4.2   Summary of fat digestion

Organ	Enzyme	Activity
Mouth	None	Mechanical mastication
Stomach	Gastric lipase (tributyrinase)	Butterfat emulsified
		Fats mechanically separated Proteins and starch digested.
Small intestine	Gall bladder—bile juice and salts (emulsifier)	Fats emulsified
	Pancreatic lipase— steapsin	Triglycerides → Mono- and diglycerides
		Triglycerides → Fatty acids
	Cholesterol esterase	Cholesterol + Fatty acids → Cholesterol esters
	Intestinal lecithinase	Lecithin → Glycerol + Fatty acid + Phosphoric acid + Choline

ABSORPTION OF LIPIDS

Lipids are not completely broken down to give fatty acids from triglycerides. The end products of digestion of fat are mono- and diglycerides, glycerol and fatty acids. The small intestine, with its millions of small villi, handles the absorption of the products of fat digestion in various ways.

The water-soluble glycerol is absorbed into the portal blood system and is carried to the liver. The fatty acids, bound to a small amount of protein travel in plasma and are absorbed into the portal vein and carried to the liver. The less polar mono- and diglycerides and very-long-chain fatty acids are transported by the bile salts which act as a ferry system.

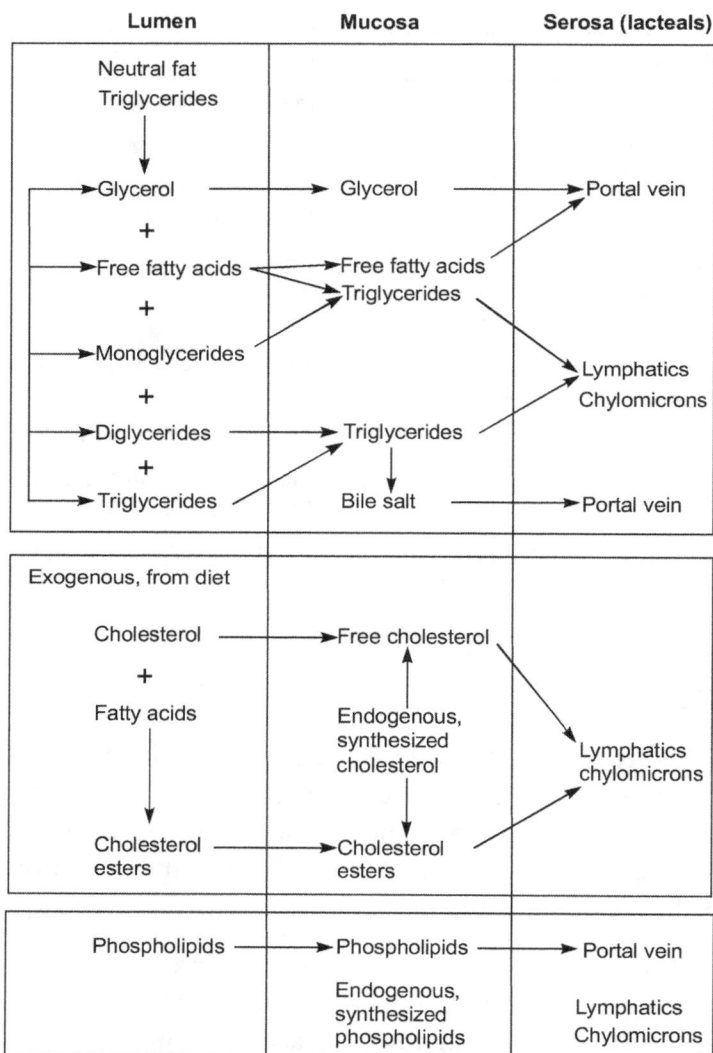

Figure 4.5   Absorption of fat, cholesterol and phospholipids

Cholesterol (from external sources) may be taken as such. Cholesterol is synthesized by the body itself. Some cholesterol (unesterified) is absorbed with the help of bile salts. Most of the cholesterol (esterified with fatty acids) is absorbed with the help of bile salts.

The phospholipid, lecithin, is absorbed as such from the diet. The other phospholipids are directly absorbed by the bloodstream, as they are hydrophilic in nature. The phospholipids are synthesized in the intestine and they form part of the lipoprotein complex (chylomicrons). The lipids are then absorbed into the portal blood circulation by the thoracic duct.

The intestinal mucosa absorbs fat and resynthesizes triglycerides and related products. The complete absorption mechanism is given in Figure 4.5.

METABOLISM OF LIPIDS

There are five factors working together to control the level of plasma lipids. They are:

    i.   diet

    ii.   synthesis of fat in the tissues

    iii. mobilization of fat from depots

    iv. rate of oxidation in various body tissues

    v.   deposition of fat in adipose tissue and the liver.

Lipids are utilized, mainly at two sites—liver and the adipose tissue. The basic interdependent relationship is called liver–adipose tissue axis.

The metabolic activities in these two sites are

i.   synthesis of lipid, lipogenesis

ii. mobilization and oxidation of lipid, lipolysis.

These two activities are so balanced that the blood lipid level is just sufficient to meet the energy needs. The summary of lipid metabolism is given in Figure 4.6.

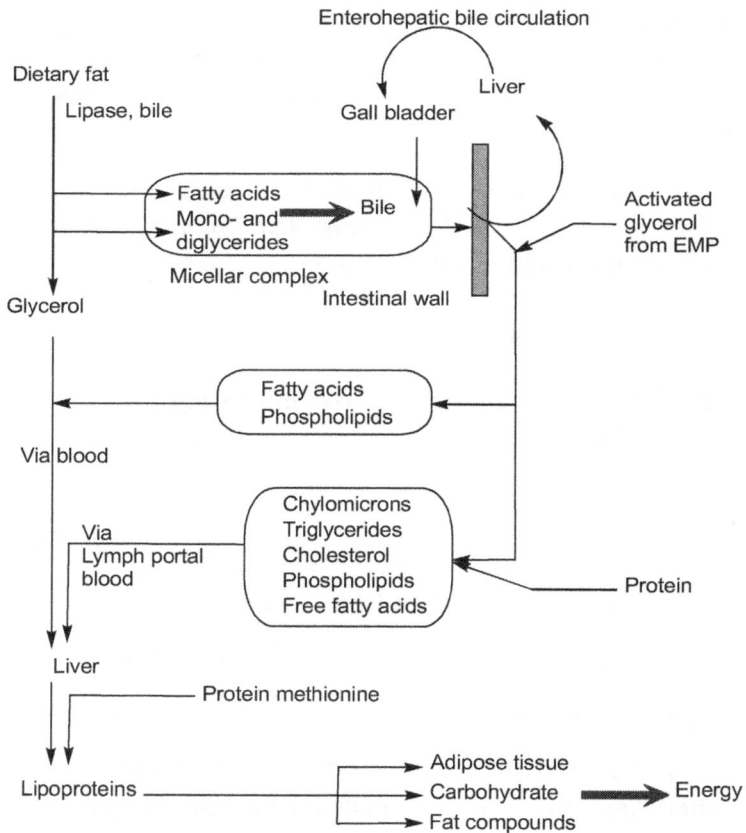

Figure 4.6   Summary of lipid metabolism

LIPID IN BLOOD

Normal human plasma contains about 500 mg of total lipid per 100 mL blood. About 120 mg triglycerides, 160 mg phospholipids and 180 mg cholesterol contribute to the total lipid content of every 100 mL blood. The two lipoproteins that are present in plasma help the transport of fat.

The normal cholesterol level in blood (180 mg/100 mL) is always safe. Higher cholesterol level (> 250 mg/100 mL) is risky for a person as it leads to coronary heart disease, hypercholesterolaemia. This trend is influenced by the following factors:

※　calorie intake

※　cholesterol intake

※　fat intake

※　essential fatty acid content in fat.

ANALYSIS OF LIPIDS

Lipids can be analysed both qualitatively and quantitatively. Lipid samples are prepared for analysis via melting the solid fats followed by filtration using hot-water funnel. Oils that are not clear must be used after filtration. The principles (not the elaborate techniques) involved in these analyses are discussed here.

Qualitative Analysis

Analysis of moisture content and volatile matter
A definite quantity of the sample is taken in an aluminium dish. It is dried in a vacuum oven at 20–25°C. The sample is weighed for constant weight.

Measurement of specific gravity   The sample is taken in a pycnometer, which is maintained at a constant temperature. It is weighed for a constant weight. The specific gravity is determined using the following equation.

$$\text{Sp. gravity} = \{W_{\text{oil-filled pycnometer}} - W_{\text{empty pycnometer}}\}/W_{\text{water}}$$

Refractive index   The refractive index is measured at 25°C for oils and at 40°C for fats. Then, the refractive index at standard temperature is calculated using the following formula.

$$R = R_1 + K\,(T_I - T_s)$$

where,

K is correction factor = 0.55 (oil) or 0.58 (fat)

R_I is refractive index at temperature T_1.

Melting/boiling point   The melting point or boiling point of the oil or fat is measured using an electrical instrument or an oil bath.

Quantitative Analysis

Total fat content   A known weight of the sample (Wg) is taken in an extraction flask. Appropriate volumes of ethanol, ether and petether solvents are added. The mixture is shaken thoroughly and gently and allowed to stand for some time. The organic layer separated is taken out in a preweighed (W_1 g) aluminium crucible. The solvents are evaporated by heating over a steam bath. The dish is cooled and weighed (W_2 g) again. The difference in weights gives the fat content that is extracted.

$$\text{Percentage fat content} = [(W_2 - W_1)/W] \times 100$$

Saponification value It is the number of milligrams of KOH required to saponify one gram of fat or lipid. It is an index of mean molecular mass of the glycerides. To a definite amount of oil or fat that is taken, more than twice the amount of KOH is added. The excess, unreacted KOH is titrated with a standard acid. A blank titration is also carried out.

From the titration value, the saponification value (S.V.) or saponification equivalent (S.E.) is calculated.

$$S.V. = \frac{(T_b - T_s) \times (N_A) 56.1}{W_{oil}} \quad S.E. = \frac{3000(W_{oil})}{(T_b - T_s) \times (N_A)}$$

Iodine value It is the number of milligrams of iodine reacted with one gram of fat. A known amount of oil or fat is treated with a known surplus amount of KI solution. The liberated iodine is titrated with standard thio. A blank is also carried out. From the titration values, the iodine value (I.V.) is calculated.

$$I.V. = \frac{(T_b - T_s) \times (N_{thio}) 12.69}{W_{oil}}$$

REVIEW QUESTIONS

Give short answers

1. Define lipid.

2. Name the organic acid present in butter. Draw its structure.

3. What is fatty acid? Give the total number of carbons and the number of unsaturation points in the chain.

4. Give the chemistry of rancidity.

5. What do you understand from the term "saponification"?

6. Lecithin belongs to the class _____. Draw its structure.

7. What is atherosclerosis?

8. Give the approximate amount of lipid content in 100 mL of the blood.

9. What are the factors that influence hypercholaesterolaemia?

10. Brief the principle of measurement of specific gravity of lipid.

11. Define saponification equivalent.

12. What is iodine value?

13. Cholesterol belongs to the class _____. Draw its structure.

14. How is oleic acid different from linoleic acid by structure?

15. Why are vegetable oils mostly liquids while animal fat are solids?

Give detailed answers

1. How are lipids classified? Give one example for each type of lipid.

2. Draw the chart of food sources that range from saturated fat to unsaturated fat.

3. List the fatty acids having aliphatic carbon chains with

    i. $C_{10} - C_{20}$ (saturated)

    ii. $C_{18} - C_{20}$ (unsaturated)

4. Write the reactions of a triglyceride with each of the following reagents:

   i. H_3O^+

   ii. NaOH

   iii. H_2, $CuCr_2O_4$

   iv. H_2, Ni

   v. O_2, light

5. Give two examples each of phospholipid and sphingolipid.

6. Discuss the biological functions of lipids.

7. Give the summary of digestion of fat.

8. Explain, in short, the metabolism of lipid.

9. How are lipids estimated quantitatively by the following methods?

   i. Saponification method

   ii. Iodination method

10. Match the following:

1. Capric	a. Synthesized with the help of lipids
2. Waxes	b. Unsaturated fatty acid (olive oil)
3. Oleic acid	c. Compound lipid (sphingolipid)
4. Sphingomyelin	d. Has no nutritional source
5. Hormones	e. Fatty acids with C_{10}

PROBLEMS

1. Exactly 1.971 g of a food sample is extracted with petether using a Soxhlet apparatus. Then, the ether and moisture contents are evaporated by keeping in a vacuum desiccator. The fat content present in the petether extract is 0.00688 g. Find the percentage of fat.

   Ans: 0.3495

2. A food sample (0.1663) is dissolved in CCl_4. To this solution a known quantity of IBr is added. It is kept in an airtight bottle for 30 minutes. A definite amount of KI is added to the mixture in the bottle. Then, it is titrated with standard thio (0.0498 N). The titration value is 13.3 mL. A blank titration is also carried out (16.8 mL). Calculate the iodine value of the sample.

   Ans: 13.3

3. An oil sample is dissolved in CCl_4 and a known quantity of IBr solution is added. This mixture is kept in an airtight bottle. After 30 minutes, a definite amount of KI is added to this mixture. The whole content is titrated with a standard thio (0.04851N). The titration value is 11.2 mL. The blank titration value is 15.3 mL. What is the amount of oil sample when the iodine value is 14.5?

   Ans: 0.1741 g

4. Accurately 0.1620 g of an oil sample is treated with 20 mL of (N/2) alcoholic KOH. Then it is cooled and titrated with 0.501N HCl using phenolphthalein indicator. The titration value is 7.8 mL. A blank titration is done without oil (8.0 mL). Calculate the saponification value.

   Ans: 312.29

5. A known quantity of oil is dissolved in 25 mL of (N/2) alcoholic KOH. It is heated in a water bath for 30 minutes and cooled. The content is titrated with 0.4976N HCl using phenolphthalein indicator. The titration value is 15.4 mL. The blank titration value is 25.5 mL. The saponification value is 184.3. Find out the amount of the oil sample.

Ans: 1.529 g

VITAMINS

INTRODUCTION

Vitamins are one of the nutrients of food. The term "vitamin" is derived from the term *vital-amine* as it was identified to be a nitrogen-containing compound that was vital for life according to the chemist Casimir Funk (Lister Institute, London). He found out a compound in the year 1911.

Vitamins have two important characteristics:

1. They are vital organic dietary substances other than carbohydrates, proteins and lipids, that are required in small quantity for a particular metabolic function.

2. They are supplied through food, as they are not synthesized by the body. Deficiency of vitamins causes biological disorders in the body.

CLASSIFICATION OF VITAMINS

Vitamins are classified into two types based on the solubility in water. Some vitamins are soluble in water (**water-soluble vitamins**). Some other vitamins are soluble only in non-polar solvents (**fat-soluble vitamins**). The water-soluble vitamins are vitamin C and vitamin B-complex. The fat-soluble vitamins are vitamins A, D, E and vitamin K. Table 5.1 gives the list of vitamins and their nutritional sources. There is one special category of vitamin that is neither water-soluble nor fat-soluble. It is vitamin H, about which not much information is available. The chemistry of some vitamins is discussed in this chapter.

VITAMIN A

The diterpenoid unit of vitamin A is known as **retinol** or **axerophthol**. E.V. McCollum and co-workers of Johns Hopkins University, Baltimore, identified it first in 1917. It is very important for eyesight or vision.

PHYSICAL AND SPECTRAL CHARACTERISTICS

It was originally isolated as a yellow oil. But, later Baxter *et al.* obtained it as a crystal (m.p. 63–64°C). It is fat-soluble substance. It gives blue colour with antimony chloride in chloroform (**Corr–Price reaction**). It exists in two isomeric forms—A_1 (*cis*) and A_2 (*trans*). It absorbs light of wavelength 325 nm (ε, 51000). It is optically active and resistant to heat.

CHEMICAL CHARACTERISTICS

Carotenoids are the precursors of vitamin A. Unsubstituted β-carotene is the major active component of vitamin A, while α and γ-carotenes are less active components of vitamin A.

Some of the chemical reactions of vitamin A are outlined here. Most of the reactions are photochemical in nature. Retinol,

Table 5.1 Some of the vitamins and their nutritional sources

Type	Vitamin	Nutritional sources
Fat-soluble vitamins	Vitamin A	Cream, butter, egg yolk, liver, carrot, sweet potato, apple, apricot, spinach, collard, broccoli, cabbage, etc.
	Vitamin D	Yeast, fish-liver oil, milk, etc.
	Vitamin E	Vegetable oil, milk, eggs, muscle meat, fish, cereals, leafy vegetables, etc.
	Vitamin K	Cabbage, cauliflower, greens, vegetables, kale, tomato, cheese, liver, etc.
Water-soluble vitamins	Thiamine (Vitamin B_1)	Pork, beef, liver, grains, legumes, egg, fish, some vegetables, etc.
	Riboflavin (Vitamin B_2)	Lactoflavin (milk), liver, kidney, some vegetables, cereals, etc.
	Niacin	Grains, corn, rice, some vegetables, etc.
	Pyridoxine (Vitamin B_6)	Yeast, wheat, corn, liver, kidney, milk, egg, some vegetables, etc.
	Pantothenic acid	Liver, kidney, egg yolk, milk, cheese, legumes, yellow corn, etc.
	Biotin	Egg yolk, liver, kidney, tomato, yeast, etc.
	Folic acid	Liver, kidney, green leafy vegetables, fruits, poultry, etc.
	Cyanocobalamine (Vitamin B_{12})	Liver, kidney, meat, egg yolk, cheese, etc.

on oxidation with oxygen, yields retinal and retinoic acid. Irradiation of all *trans*-vitamin A in hexane, yields a mixture of *trans*-, 13-*cis*, 9-*cis* isomers, in the proportion 50%, 45% and 8% respectively.

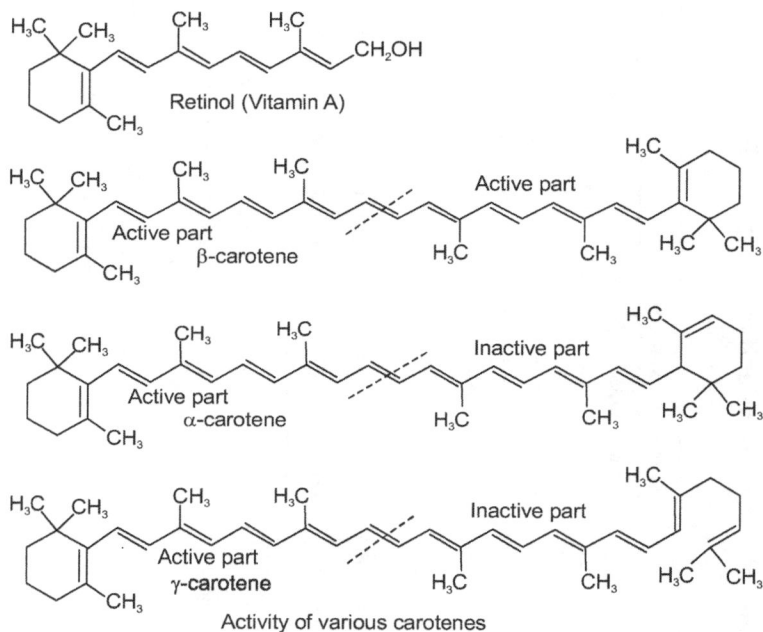

Activity of various carotenes

Figure 5.1 illustrates the various reactions of vitamin A. On photochemical dimerization, vitamin A gets dimerized to give a polyene type of compound (I). Vitamin A undergoes photooxidation in presence of and a sensitizer to give polyene heterocyclic compounds (II and III). Thermal degradation of vitamin A yields ketone type of compounds (IV and V). Vitamin A gives hydroxy ether type of compound (VI) on "ene" reaction, followed by hydrolysis.

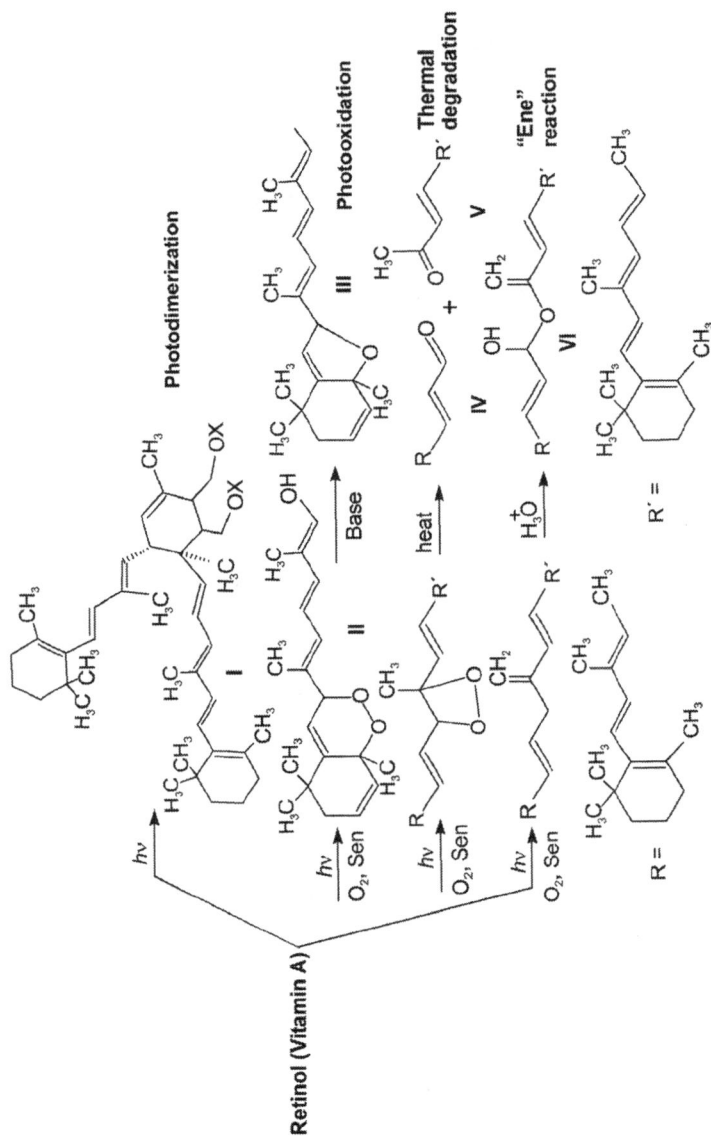

Figure 5.1   Structure and reactions of vitamin A

BIOLOGICAL FUNCTIONS

❊   Vitamin A plays an important role in the development and growth.

❊   It regulates and enhances the stability of biological membranes.

❊   It maintains mucus-secreting cells of epithelia.

❊   It helps biosynthesis of glycoproteins.

❊   It prevents keratinization.

❊   It plays a very active role in vision (eyesight).

❊   It helps the formation of some epithelial tissues, which are useful in the synthesis of teeth enamel.

ABSORPTION OF VITAMIN A

Absorption of vitamin A is aided by bile salt, pancreatic lipase and some fat. Oxygen easily destroys vitamin A. Bile salt acts as a natural antioxidant. Also, it serves as a vehicle of transport through the intestinal wall. Pancreatic lipase helps splitting the lipid and forming an emulsion of an aqueous dispersion form of vitamin A. This is easy for absorption.

DEFICIENCY DISORDERS

When the normal requirement of vitamin A is not supplied, it leads to following deficiency disorders.

❊   Xerophthalmia

❊   Night blindness

❊   Keratinization of epithelium

❊   Follicular hyperkeratosis

❊   Skin and mucous membrane infections

❊   Faulty tooth formation

RECOMMENDED DAILY ALLOWANCE

The daily requirements of vitamin A, as recommended by the National Research Council of USA are given here. For children it varies from 1500 up to 5000 International Units (I.U.). Male adults need about 5000 I.U. and female adults need 4000 I.U. For pregnant and lactating mothers, the requirement is about 5000 and 6000 I.U. respectively.

VITAMIN D

It represents a group of closely related five fat-soluble vitamins—D_1, D_2, D_3, D_4 and D_5. They are structurally related to sterols. When sterols are irradiated with UV light, these vitamins are produced. Mellanby discovered this vitamin in the year 1919. It was isolated in the crystalline form as early as 1930.

PHYSICAL CHARACTERISTICS AND CHEMICAL CHARACTERISTICS

It is a colourless crystal, stable to heat and alkali and is not easily oxidized. It is less polar and soluble in less polar solvents. It is insoluble in water. The provitamin D undergoes ring-opening reaction of the cyclohexadiene ring (B-ring) on irradiation. The product undergoes [1,3] sigmatropic reaction to form vitamin D. The reactions are depicted in Figure 5.2.

Figure 5.2 depicts the chemical reactions of vitamin D_2. The provitamin D_2 undergoes photochemical conversion to vitamin D_2. In the same way provitamin D undergoes photochemical conversion to vitamin D.

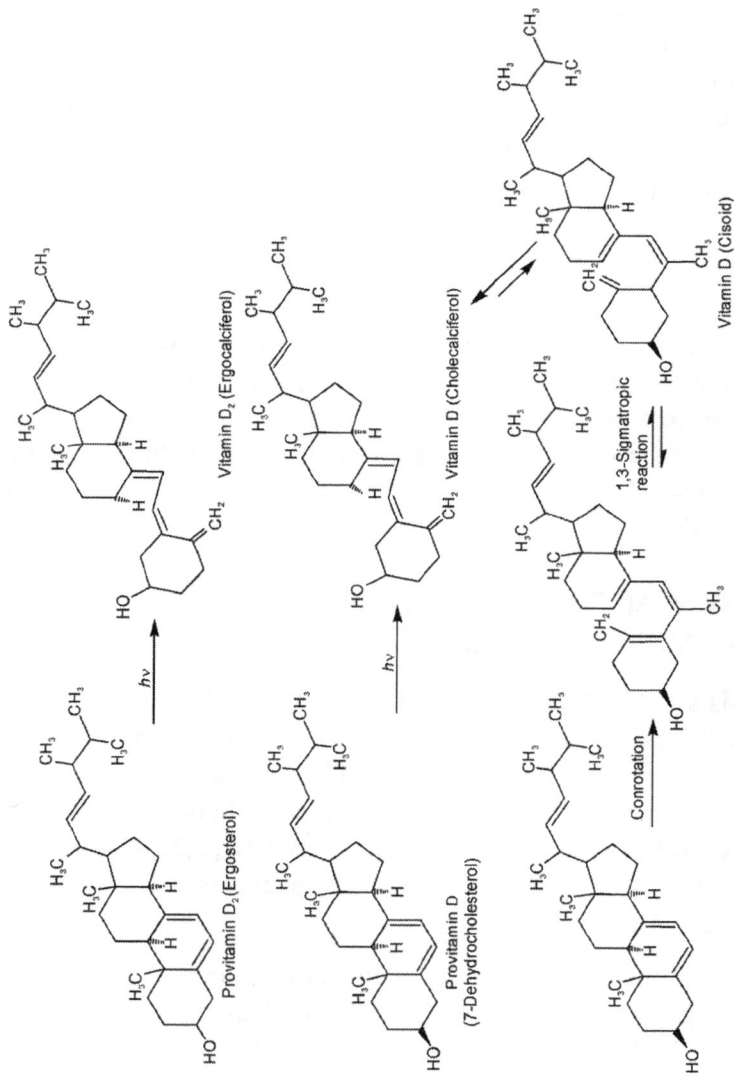

Figure 5.2   Reactions of vitamin D

BIOLOGICAL FUNCTIONS OF VITAMIN D

❋ Absorption of Ca and P in the intestine making use of the vitamin D-dependent calcium-binding protein.

❋ Mobilization of Ca and P from the bone and bone growth by the appropriate supply of Ca and P.

❋ Renal readsorption of Ca and P.

❋ It plays a very active role in citrate metabolism.

ABSORPTION OF VITAMIN D

Absorption of vitamin D is always accompanied by the absorption of Ca and P in the small intestine. Absorption of vitamin D requires the presence of bile salts. Vitamin D (like vitamin A) is absorbed by mineral oil. Therefore, if mineral oil is taken, it should be ingested separately from food. Vitamin D is absorbed through skin and carried to the liver and other organs for use. It is stored in liver and in body's fatty tissues. Vitamin D is excreted from the circulating blood by way of the bile salt.

DEFICIENCY DISORDERS

❋ Rickets is a deficiency disease indicated by the bulging of the forehead, and by the bone becoming soft and fragile. Wrists, knees and ankle joints get enlarged. Restlessness and nervous irritability result.

❋ Tetany is another disease that is characterized by low serum calcium, muscle twitching, cramps and convulsions.

❋ Dental disorders like tooth decay and malformation of teeth.

✳ Osteomalacia is the softening of bone (adult rickets) and characterized by pain in the bones and spontaneous multiple fractures.

RECOMMENDED DAILY ALLOWANCE

The National Research Council of US recommends 400 I.U. daily for children and for lactating mothers. There is no fixed amount for the adults. It is generally advised that exposure to sunlight should be sufficient. One I.U. of vitamin D is equivalent to the biological activity of 0.025 µg of pure crystalline vitamin D_3.

VITAMIN E

Evans and Bishop established the fact that a fat-soluble chemical factor was much necessary for the reproduction in rats. So, they named it **tocopherol**. The two forms of vitamin E in varying degrees of activity are **tocopherol** and **tocotrienol**. But, the most active form of vitamin E is α**-tocopherol**.

PHYSICAL AND CHEMICAL CHARACTERISTICS

Vitamin E is resistant to heat and acids but is decomposed by UV light. It is somewhat a polar substance and is soluble in slightly polar solvents. Tocopherol and tocotrienol are structurally related to chromone type of compounds. Many structural forms and some reactions are given in the following section.

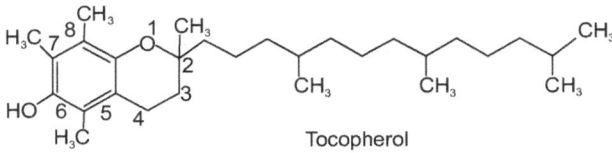

Tocopherol

Substitution	Tocopherol
5,7,8-trimethyl	α-tocopherol
5,8-dimethyl	β-tocopherol
7,8-dimethyl	γ-tocopherol
8-methyl	δ-tocopherol

Figure 5.3 illustrates the chemical reactions of vitamin E. It gets oxidized in presence of silver nitrite to give α-tocopherol quinone. Another oxidation of vitamin E with ferric salt, vitamin C and oxygen yields α-tocopherol epoxide. Vitamin E on pyrolysis gives hydroxyquinone and an unsaturated hydrocarbon of C_{19}.

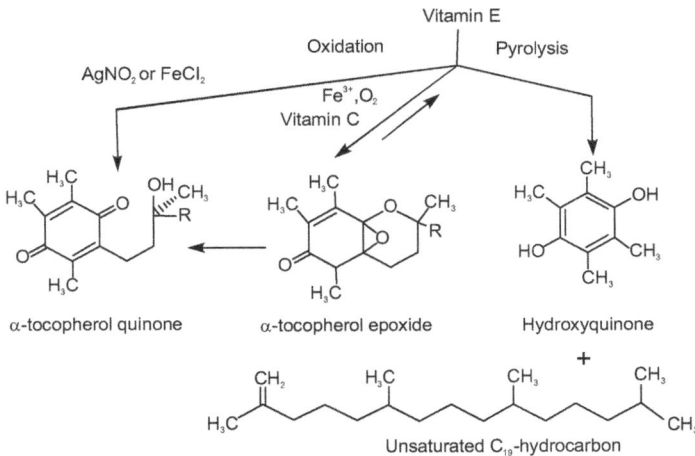

Figure 5.3   Reactions of vitamin E

Tocopherols are very efficient in quenching singlet oxygen by a combination of chemical and physical quenching processes:

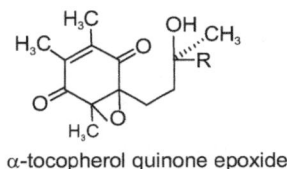

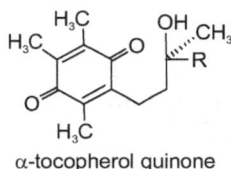

α-tocopherol quinone epoxide   α-tocopherol quinone

BIOLOGICAL FUNCTIONS

❊   It plays an important role in the reproductive activity in animals.

❊   It is useful for muscular integrity. It maintains the structure and functions of smooth muscle, skeletal muscle, cardiac muscle, vascular tissue, etc.

❊   It is important for the liver integrity.

❊   It protects the red blood cells by inhibiting the oxidase action on haemoglobin.

❊   It acts as a cofactor in various enzyme systems involved in the respiration and in the biosynthesis of cellular substances like DNA.

❊   It acts as an electron-transferring agent in the cell's energy metabolism.

ABSORPTION OF VITAMIN E

It is absorbed like the other fat-soluble vitamins through bile salts and lipids. It is stored in different tissues, especially in adipose tissue. Malabsorption of vitamin E causes muscle defects. Vitamin E is proved to have a direct correlation with protection of unsaturated fatty acids.

DEFICIENCY DISORDERS

❋ Reproductive failure

❋ Macrocytic anaemia

❋ Shorter lifespan of red blood cells

❋ Liver necrosis

❋ Muscular dystrophy

RECOMMENDED DAILY ALLOWANCE

The biochemical mechanism of vitamin E in different biological functions is not exactly known. But, it is a very important nutrient. Requirements vary with the amount of polyunsaturated fatty acids present in the diet. The recommended daily allowance by the National Research Council, US, are given here. Children in the age group of 0.5 to 10 years may require 4–10 I.U. of vitamin E. Males in the age group 11 and 51+ may need 12–15 I.U. and the females require in the range 12–15 I.U. Lactating mothers need 15 I.U.

VITAMIN K

Vitamin K was identified as a *Koagulations* vitamin by Professor Henrik Dam (Biochemistry, University of Copenhagen) in the year 1929. He found that chicks showed haemorrhagic disease when fed with fat-free diet due to the absence of a factor responsible for the clotting of blood. He got the Nobel Prize in 1943 for this work.

There are three types of vitamin K. Two of them K_1 (**phylloquinone or phytonadione**) and K_2 (**farnoquinone**) occur in nature. **Menadione** is purely synthetic; it is the soluble synthetic form of vitamin K_3. Vitamin K is synthesized by the bacteria of the intestine. So, there is always adequate supply of vitamin K in the body.

PHYSICAL AND SPECTRAL CHARACTERISTICS

It is fat-soluble and resistant to heat. It is destroyed by acids, alkalies, and oxidizing agents, and by light. Vitamin K_1 is a yellow oil whereas vitamin K_2 is a yellow solid (m.p. 54°C). Both show absorptions at wavelengths 243, 249, 260 and 270 nm. (ε = 20,000) and at 325 nm (ε = 3000). These absorptions are due to the chromophoric groups present.

CHEMICAL CHARACTERISTICS

The chemical structures are given in the following figure. They have the basic skeleton 2,3-disubstituted naphthoquinones. The difference is only in the side chain.

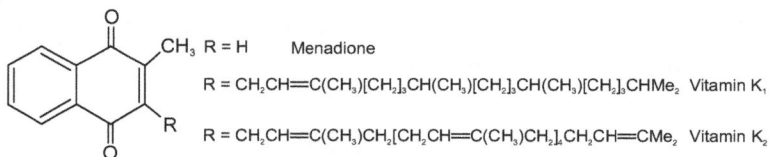

R = H   Menadione

R = $CH_2CH=C(CH_3)[CH_2]_3CH(CH_3)[CH_2]_3CH(CH_3)[CH_2]_3CHMe_2$ Vitamin K_1

R = $CH_2CH=C(CH_3)CH_2[CH_2CH=C(CH_3)CH_2]_5CH_2CH=CMe_2$ Vitamin K_2

BIOLOGICAL FUNCTIONS

※ It catalyses the synthesis of prothrombin by the liver. It initiates blood clotting.

※ It acts as an essential factor in the oxidative phosphorylation.

※ It is required for the synthesis of other proteins containing γ-carboxyglutamic acid.

ABSORPTION OF VITAMIN K

Bile salts are required for the absorption of vitamin K. They are absorbed along with other fat-related products by way of abdominal lacteals, lymphatic systems, portal blood and then to liver. It is stored in small amounts. An overdose of this vitamin is excreted.

DEFICIENCY DISORDERS

✳ Deficiency of vitamin K leads to low blood level of prothrombin and other clotting factors. This results in haemorrhage.

✳ Bleeding tendencies in urinary diseases or surgical procedures.

✳ Intestinal malabsorption leading to diseases such as celiac and sprue.

RECOMMENDED DAILY ALLOWANCE

No definite amount is recommended as the daily requirement. The deficiency is unlikely except in the clinical situation. Intestinal bacteria constantly synthesize and supply this vitamin. The amount of vitamin the body needs is apparently small. The liver produces prothrombin if vitamin K is to be effective.

VITAMIN B-COMPLEX

It is a fat-soluble vitamin. In 1912, **Casimir Funk** found out that this substance was very effective in preventing beriberi. E.V. McCollum termed this vitamin as a water-soluble B. Vitamin B is not a single substance. It is a group of compounds, which we designate as **vitamin B-complex**. They possess very important biological functions as they form part of various enzyme systems. This vitamin B-complex includes the following compounds:

1. Thiamine (B_1)
2. Riboflavin (B_2)
3. Pantothenic acid (B_5)
4. Folic acid (B_9)
5. Pyridoxine (B_6)
6. Niacin (B_3)
7. Cyanocobalamine (B_{12})

VITAMIN B_1: THIAMINE

It is otherwise termed anti-beriberi factor. It occurs in yeast, milk, groundnut, egg and outer cover of rice, wheat, etc. Thiamine pyrophosphate that forms part of cocarboxylase is a biologically active form.

PHYSICAL AND CHEMICAL CHARACTERISTICS

It is a water-soluble vitamin. It is crystallized as a white crystalline solid. It is pyridine hydrochloride linked with a thiazole. Some of the chemical characteristics are given in Figure 5.4.

Vitamin B_1 reacts with bisulphite to form the sulphite salt of 2-methyl 4-aminopyrimidine and 4-methyl 5-(β-hydroxyethyl) thiazole. Acid reacts with vitamin B_1 producing 2-methyl 4-amino 5-hydroxymethyl pyrimidine and 4-methyl (β-hydroxyethyl)thiazole. Vitamin B_1 reacts with base maintained at pH 11, forms dihydrothiochrome, that gets degraded further to give thioketone and 2-methyl 4-amino 5-aminomethylpyrimidine. Vitamin B_1 on oxidation gives thiochrome. Photochemical degradation of vitamin B_1 yields ketothiol type of compound.

BIOLOGICAL FUNCTIONS

❊ It takes part in the carbohydrate metabolism.

❊ It is a cofactor for **transketolase** and it helps produce active glyceraldehyde.

❊ It activates the functioning of some nerve cell membranes. It influences the action of neurotransmitters—acetylcholine or serotonin.

DEFICIENCY DISORDERS

❊ Deficiency of vitamin B_1 may cause gastrointestinal disturbances, that can have many manifestations—anorexia, indigestion, constipation, gastric atony, deficient secretion of HCl, etc.

❊ Impairment of neural activity. Diminished reflex responses.

❊ Degeneration of myelin sheaths. It may cause nerve irritation.

❊ Weakening of heart muscles, tachycardia, palpitation, etc.

RECOMMENDED DAILY ALLOWANCE

It depends on the carbohydrate and energy metabolism. Various studies revealed that infants between the age group 0.5 and 3 years, require 0.7 mg. Children in the age group 4–10 years may need 0.9–12 mg. Males between 11–55 years are advised to take about 1.4 mg. Females in the age group 11–55 years are recommended about 1.2 mg. Pregnant and lactating mothers are advised to take as low as 0.3 mg.

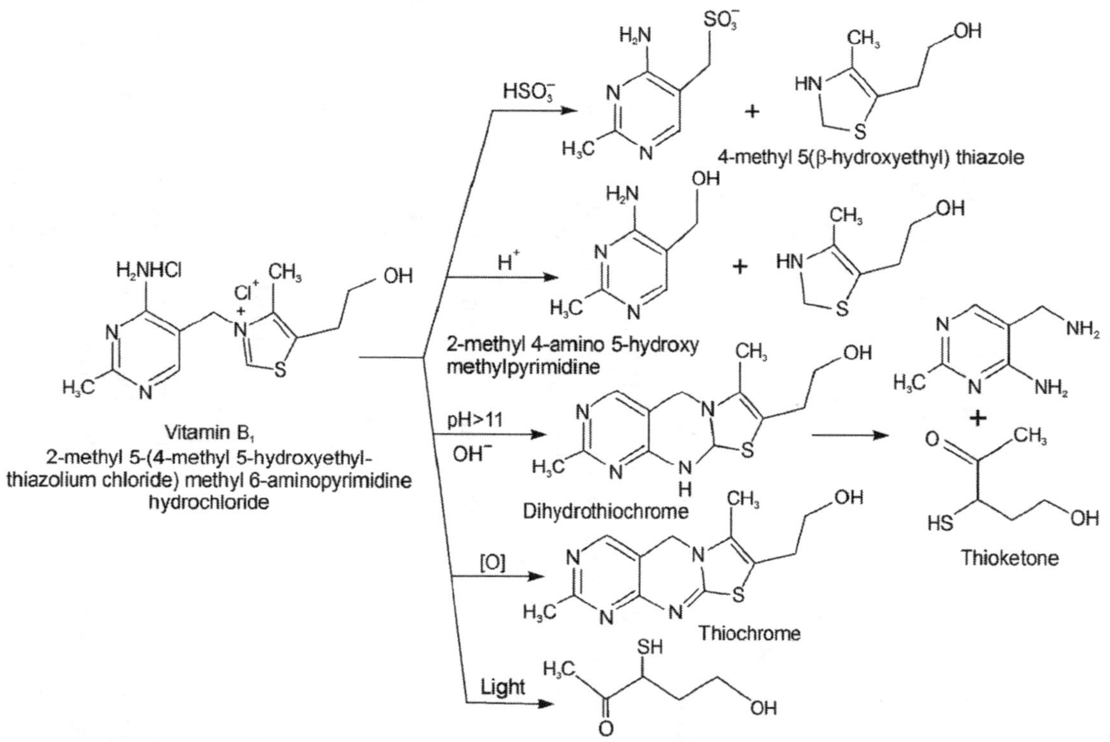

Figure 5.4 Reactions of vitamin B_1

VITAMIN B₂: RIBOFLAVIN

As early as 1897 the British Chemist **Blythe** observed the yellow pigment of milk whey. Later, in 1932 the German Chemist **Warburg** discovered vitamin B_2 as the yellow substance containing the pentose sugar to be present in the yeast. Thus, it was named Riboflavin ("ribo"— sugar; "flavus"— yellow).

PHYSICAL AND SPECTRAL CHARACTERISTICS

It is an orange-yellow crystalline solid, soluble in water. It is stable to heat. Light and irradiation may destroy this substance. In aqueous solution it shows yellow-green fluorescence having the λ_{max} = 565 nm.

CHEMICAL CHARACTERISTICS

Some of the reactions of riboflavin are shown below:

Figure 5.5 depicts the chemical reactions of vitamin B_2 (flavin). Alkaline hydrolysis of flavin gives diamide and 6,7-dimethyl 2-keto 1-D-ribityl 3-quinoxaline carboxylic acid. With sodium bisulphite it gives N-hydroflavin sulphite. The photochemical excitation gives the triplet excited state of the flavin, which further changes to give deuteroflavin. The triplet excited flavin shows photosensitized oxidation of amino acids to gives ammonia, the corresponding aldehyde and carbon dioxide. Flavin undergoes photochemical dealkylation to give lumichrome.

BIOLOGICAL FUNCTIONS

＊ It acts as a coenzyme in a series of enzymatic reactions involving the enzymes like xanthine oxidase, D- and

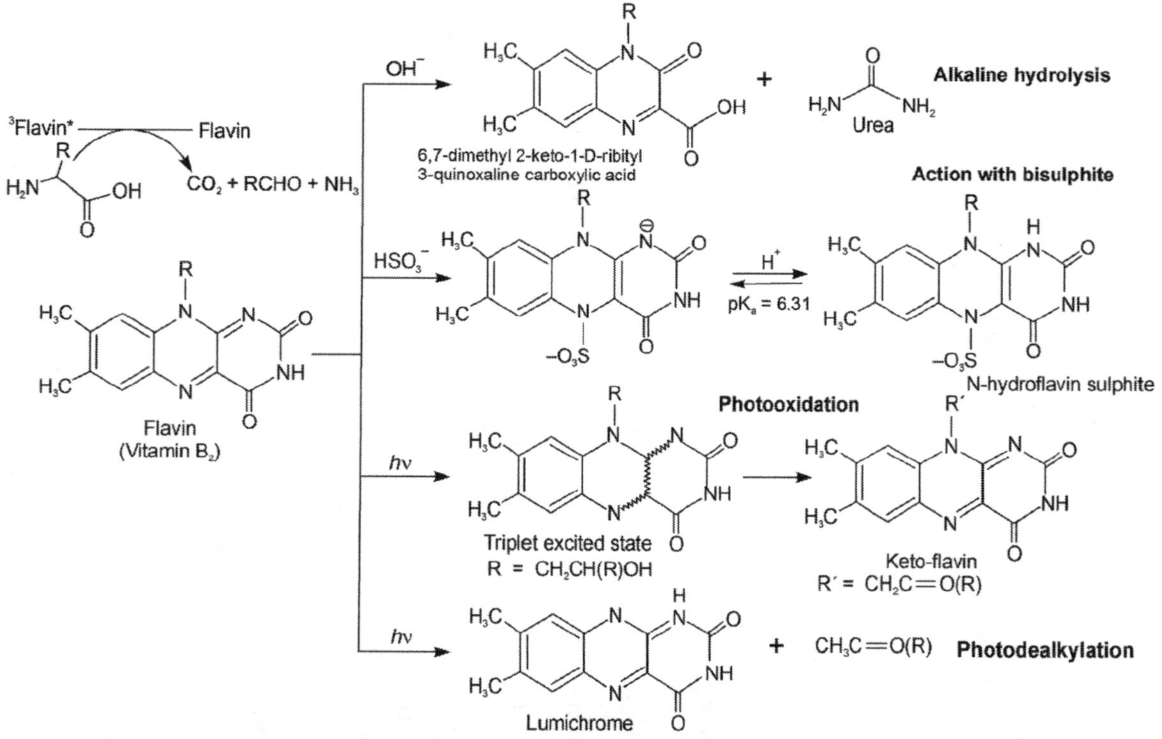

Figure 5.5 Reactions of vitamin B$_2$

L-amino oxidases, aldehydeoxidase, succinic dehydrogenase, cytochrome-*c* reductase, etc.

✳ It plays a very important role in the metabolism of carbohydrates, fats and proteins.

DEFICIENCY DISORDERS

✳ A symptom of cracks on the lips and on the mouth edges—ariboflavinosis or cheilosis.

✳ Inflammation of the tongue or glossitis.

✳ Pellagra in human beings and curled toe in chicks.

✳ Irritation of the eyes.

✳ Seborrhoeic dermatitis—scaly, greasy eruption of the skin.

RECOMMENDED DAILY ALLOWANCE

Children in the age group 1–10 years may require about 1.2 mg. For males, the amount recommended is in the range 1.5–8 mg. Females need about 1.4 mg. The pregnant and lactating women may require about 1.8 mg per day.

VITAMIN B$_5$: PANTOTHENIC ACID

It was isolated and synthesized by R.J. William in 1938. The Greek word *pantothen* means "in every corner or from all sides". Intestinal bacteria synthesize considerable amount of this compound. Deficiency of this vitamin is unlikely.

PHYSICAL AND CHEMICAL CHARACTERISTICS

It is a pale yellow oily liquid, soluble in water. It is usually stored as a white crystalline calcium salt, calcium pantothenate. The structure of the compound is given below:

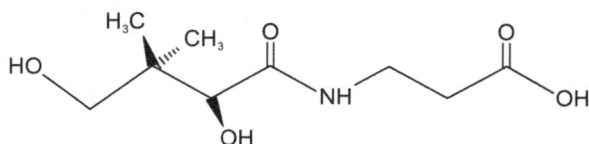

BIOLOGICAL FUNCTIONS

✳   It acts as a coenzyme in body metabolism. It plays a very important role in the activation of acetic acid, fatty acids, amino acids, etc.

✳   It is involved in the metabolism of carbohydrates and fats.

DEFICIENCY DISORDERS

✳   Burning sensation in the foot.

✳   Nausea, vomiting and irritability

RECOMMENDED DAILY ALLOWANCE

The daily requirement is not well established, as it is very much available in the body. So, any disorder due to deficiency is rare. It is generally recommended to be about 10–20 mg.

VITAMIN B_9: FOLIC ACID

Stokstad and **Manning** identified this vitamin in the year 1938 as a growth factor and termed it vitamin U. Later, in 1945 Angier, Stokstad and their team synthesized this vitamin. The name folic acid was given due to its origin, that is, the dark green leaves of spinach (L. *folium* = leaf).

PHYSICAL AND CHEMICAL CHARACTERISTICS

It is a yellow crystalline solid, soluble in water. It consists of two amino acids and one heterocyclic ring pterin. The formyl derivative of folic acid, folinic acid, is superior in activity compared to the other forms.

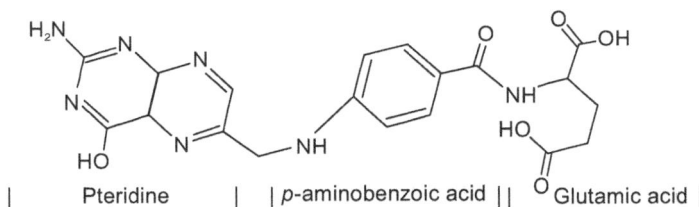

| Pteridine | p-aminobenzoic acid | Glutamic acid |

BIOLOGICAL FUNCTIONS

❈ It acts as a coenzyme for the single carbon transfer.

❈ It is useful in the formation of nucleoproteins. So it helps cell growth and reproduction.

❈ It takes part in the synthesis of thymine.

❈ It is C1-carrier in the formation of haem.

DEFICIENCY DISORDERS

❈ Nutritional megaloblastic anaemia.

❈ Sprue (gastrointestinal disease) and leukaemia.

RECOMMENDED DAILY ALLOWANCE

For the children in the age group 1–10 years, it is recommended to be in the range 50–300 µg. Males and females are advised to take 400 mg. Pregnant and lactating mothers require as high as 800 mg.

VITAMIN B₆: PYRIDOXINE

It is otherwise called antidermatitic factor; the name derived from its functionality. In 1926, **Joseph Golberger** found that this substance cured a particular dermatitis in rats. **Harris** synthesized this compound in the 1939. **Snell** and his team isolated pyridoxine from animal tissue and synthesized the two other forms, pyridoxal and pyridoxamine. It acts as a coenzyme in its phosphate forms.

PHYSICAL AND CHEMICAL CHARACTERISTICS

It is a white crystalline solid, soluble in water. It is optically active. It is stable to heat, but, sensitive to light and alkali. It is structurally related to nicotinic acid. Some of the reactions of vitamin B_6 are given in Figure 5.6.

Figure 5.6 illustrates the chemical reactions of vitamin B_6 (pyridoxal). It undergoes transamination reaction to give the ketimine form of pyridoxal(I). On β-elimination it gets unsaturated to give the compound II. Vitamin B_6 shows decarboxylation reaction to give the decarboxylated form of pyridoxal (III). Under acidic conditions, pyridoxal undergoes racemic modification.

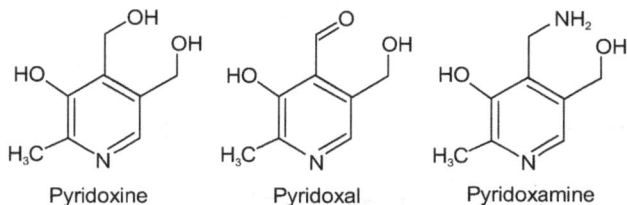

Pyridoxine   Pyridoxal   Pyridoxamine

BIOLOGICAL FUNCTIONS

※   Pyridoxal and pyridoxamine are coenzymes for enzymes like decarboxylases, transaminases and pyridoxal

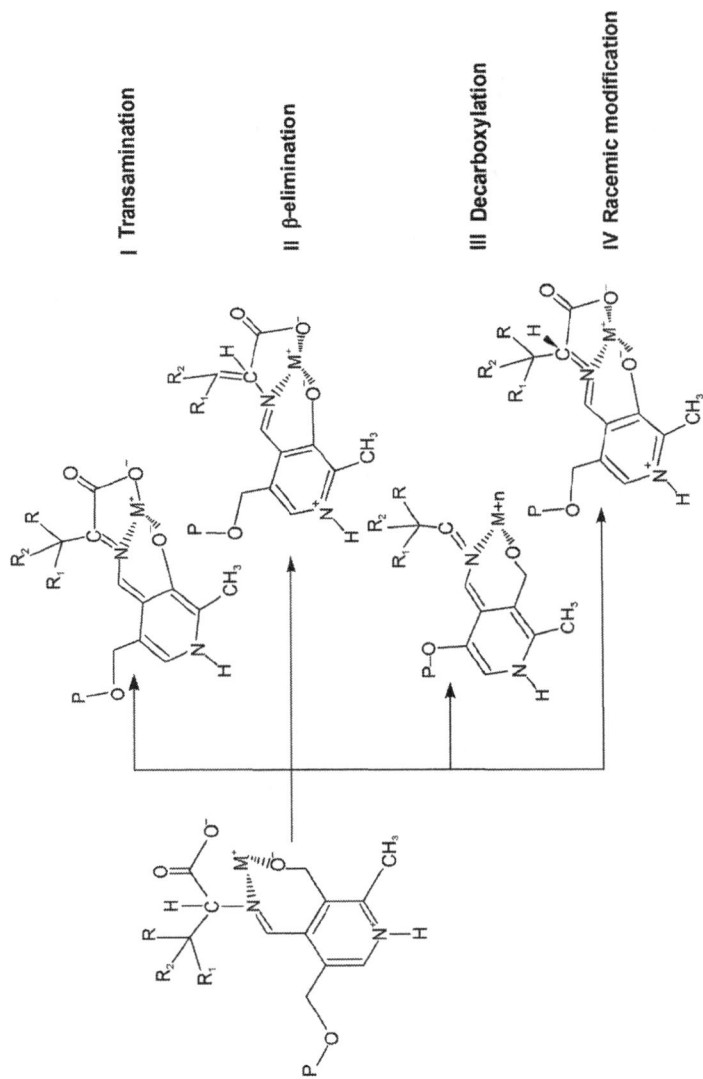

I Transamination

II β-elimination

III Decarboxylation

IV Racemic modification

Figure 5.6   Reactions of vitamin B$_6$

phosphate. (These enzymes are involved in the synthesis of tryptophan in *Neurospora crassa.*)

❋   Pyridoxine is involved in the synthesis of nicotinic acid.

❋   It helps the *trans*-sulphuration process in the formation of derivatives of cysteine.

❋   It helps incorporating the glycine and succinate (a glucose metabolite of Kreb's cycle) into haeme that forms haemoglobin.

❋   It helps conversion of the essential fatty acid, linoleic acid, to arachidonic acid.

DEFICIENCY DISORDERS

❋   A hypochromic, microcytic anaemia.

❋   Disturbances in the central nervous system.

❋   Tuberculosis.

❋   Hyperemesis of pregnancy.

RECOMMENDED DAILY ALLOWANCE

Adults require approximately about 1 mg daily. However, the National Research Council, US recommends 2 mg per day for adults. In case of infants, the requirement is fixed depending upon the individual case, as it is involved in amino acid metabolism.

NIACIN/NICOTINIC ACID

As early as the 1900s **Joseph Goldberger** discovered niacin, a cure for pellagra. In 1911, **Casimir Funk** (London) isolated nicotinic acid from the rice polishings. Later, in 1937 **Conard**

Elvehjem (University of Wisconsin) confirmed that this substance is a definite cure for pellagra and other related diseases.

PHYSICAL AND CHEMICAL CHARACTERISTICS

It is a water-soluble colourless solid. The chemical name of the compound is pyridine 3-carboxylic acid. The sodium salt and the amide of nicotinic acid are the biologically active forms. The structures are given below. The active form is nicotinamide adenine dinucleotide (NAD+). It is involved in the dehydrogenase-catalysed reactions (Figure 5.7).

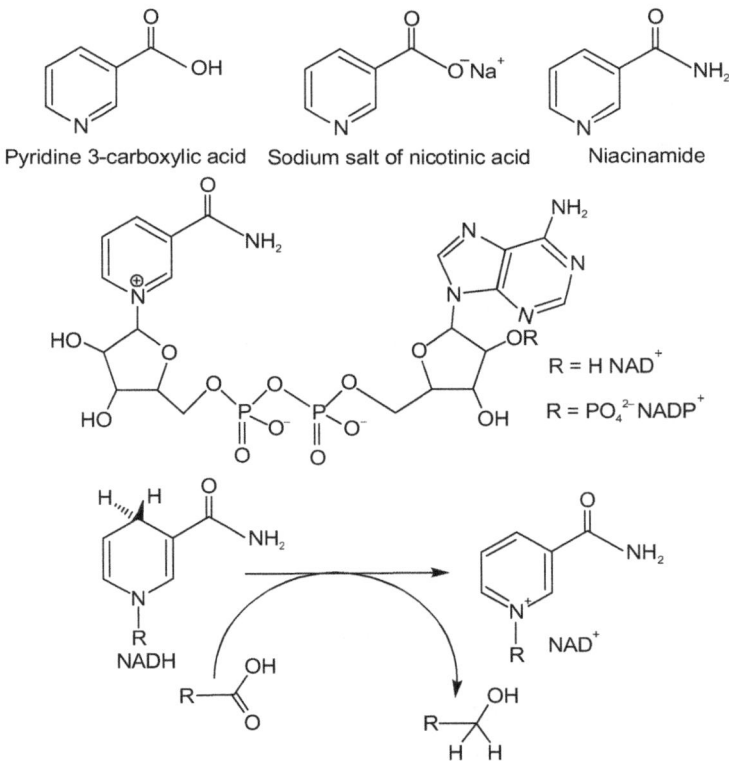

Pyridine 3-carboxylic acid   Sodium salt of nicotinic acid   Niacinamide

R = H NAD+
R = PO₄²⁻ NADP+

NADH

NAD+

Figure 5.7   Reactions of vitamin niacin

The vitamin niacin in the NADH is getting oxidized to NAD^+ by reducing a carboxylic acid to an alcohol.

BIOLOGICAL FUNCTIONS

* It forms a part of coenzymes in tissue oxidation. In cellular coenzyme systems it joins with riboflavin in order to convert proteins and fats to glucose.

* It is involved in the hydrogen transfer reactions. The enzymes involved are alcohol dehydrogenase, phosphate glyceraldehyde dehydrogenase, lactic acid dehydrogenase, malic dehydrogenase, etc.

DEFICIENCY DISORDERS

* General weakness, lassitude, anorexia, indigestion and other skin eruptions.

* A type of dark and scaly dermatitis.

* A mild malfunctioning of the central nervous system. The symptoms are confusion, apathy, disorientation and neuritis.

* Pellagra.

RECOMMENDED DAILY ALLOWANCE

The minimum for necessary tissue stores is about 9 mg per 1000 calories. Depending upon the age, growth period, pregnancy and lactation, physical activity and illness the requirements may vary. For children in the age group of 1–10 years, the amount required is 5–16 mg equivalent. Males are recommended about 18 mg equivalent. Women and pregnant women need as much as 14–16 mg equivalent.

The term "mg equivalent" means 1 mg of niacin that is derived from 60 mg of dietary tryptophan.

VITAMIN B$_{12}$: COBALAMINE

In 1948 two teams of scientists—one American and the other English—crystallized a red compound from liver. It was labelled vitamin B$_{12}$. In the same year, it was shown that the new vitamin could cure the blood-forming defect and pernicious anaemia.

PHYSICAL AND CHEMICAL CHARACTERISTICS

It is a red solid, soluble in water. It is in the complex form with cobalt ion. The cobalt metal ion forms the centre of the porphyrin nucleus to which ribose phosphate and benzimidazole are attached. The structure and some reactions of this vitamin are given below.

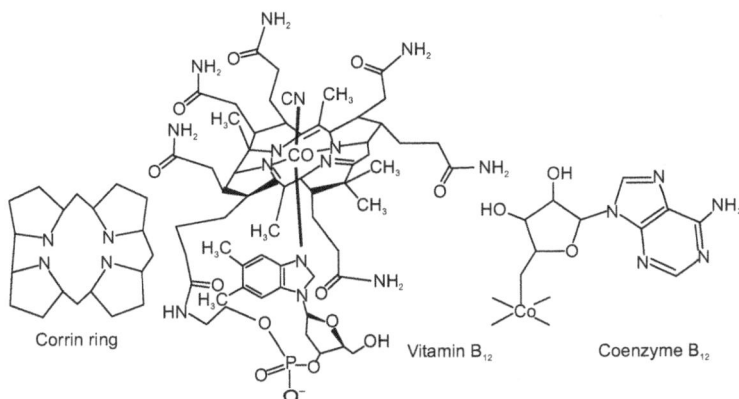

Corrin ring   Vitamin B$_{12}$ Coenzyme B$_{12}$

While irradiating with light, the Co–C bond is cleaved. Under anaerobic condition, the 5´-deoxyadenosylradical cyclizes. Under aerobic condition, the products are hydroxocobalamin and 5´-aldehyde of adenosine.

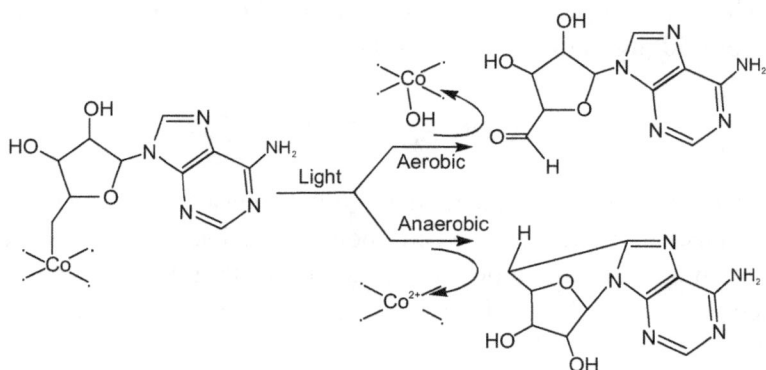

Oxidation, under mild alkaline condition, yields dehydrovitamin B_{12}. The acetamide side chain at C-7 in the corrin ring cyclizes to form a γ-lactam. With oxidizing agent like iodine, it gives γ-lactone.

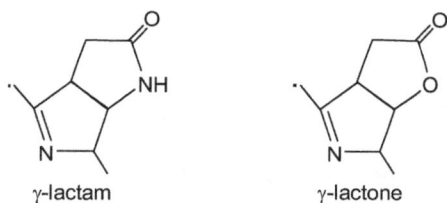

γ-lactam   γ-lactone

BIOLOGICAL FUNCTIONS

✻ It is useful as a methylating agent in many metabolic processes.

✻ It participates in the syntheses of nucleic acid and vital proteins in the cell.

✻ It takes part in the formation of red blood cells. So, it controls pernicious anaemia.

✻ It indirectly facilitates the action of folic acid. So, it helps treating "sprue", a gastrointestinal disease characterized by intestinal lesions, malabsorption defects, diarrhoea, etc.

DEFICIENCY DISORDERS

❋ Pernicious anaemia

❋ Intestinal disorder, sprue

❋ Problems in the synthesis of nucleic acid and proteins.

RECOMMENDED DAILY ALLOWANCE

The minimum requirement is in the range 0.6–1.2 µg. But, the National Research Council suggests as much as 3.0 µg for adults and about 1.5 µg for children.

VITAMIN C: ASCORBIC ACID

As early as 1500 BC the "Father of Medicine", **Hippocrates**, was aware of the disease scurvy. **Jacques Cartier** cured the dying men of his team, by giving a brew made of pine needles and bark. Norwegian scientists in 1907 reported that the food deficient of ascorbic acid could cause scurvy. In 1928, **Szent–Gyorgyi** isolated hexuronic acid derivative from cabbage, orange juice and from adrenals. In 1932, **Charles Glen King** and **W.A. Waugh** University of Pittsburg isolated hexuronic acid in lemon. They proved that it could cure scurvy.

PHYSICAL AND CHEMICAL CHARACTERISTICS

It is a white crystalline solid soluble in water. It is acidic in nature. It is L-ascorbic acid. Structurally it is related to L-glucose. D-isoascorbic acid is a common substitute for the L-ascorbic acid, that is used in most commercially prepared foods.

L-ascorbic acid is dibasic $(pK_{a_1} = 4.0$ and $pK_{a_2} = 11.3)$. The acidity of ascorbic acid is due to the dissociation of proton at 3-hydroxy group. Thereby, a monoanion is formed.

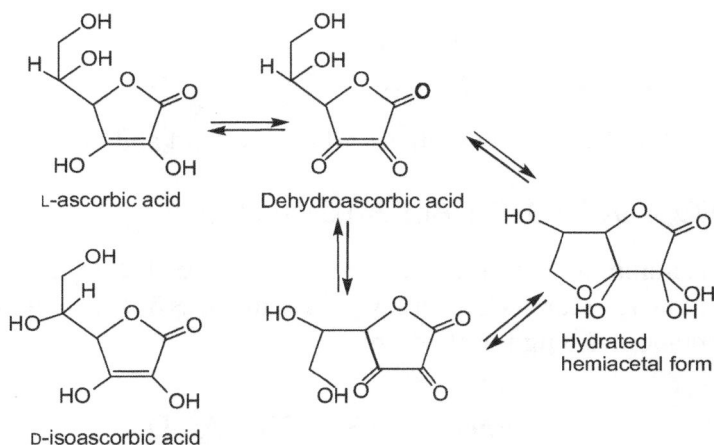

Figure 5.8 shows the synthesis of L-ascorbic acid. D-glucose is oxidized using bromine-water to form D-glucuronic acid. It is further reduced using hydrazine to give L-gluconic acid. Gluconic acid is dehydrated further to form the L-gluconolactone. It is oxidized and subsequently enolized to yield L-ascorbic acid.

Ascorbic acid reacts with a base giving a monoanion. Acid hydrolysis of the ascorbic acid by anaerobic route gives gluconic acid which further reacts to give furfuraldehyde. Oxidation of ascorbic acid gives pentosulose. Ascorbic acid reacts with metals like Fe^{3+} and Cu^{2+} in presence of oxygen to yield dehydroascorbic acid. Ascorbic acid undergoes Strecker degradation to give L-scorbamic acid and further to give 2,2′nitrilodi-2(2′)-deoxy-L-ascorbic acid ammonium salt. Figure 5.9 shows the reactions of L-ascorbic acid.

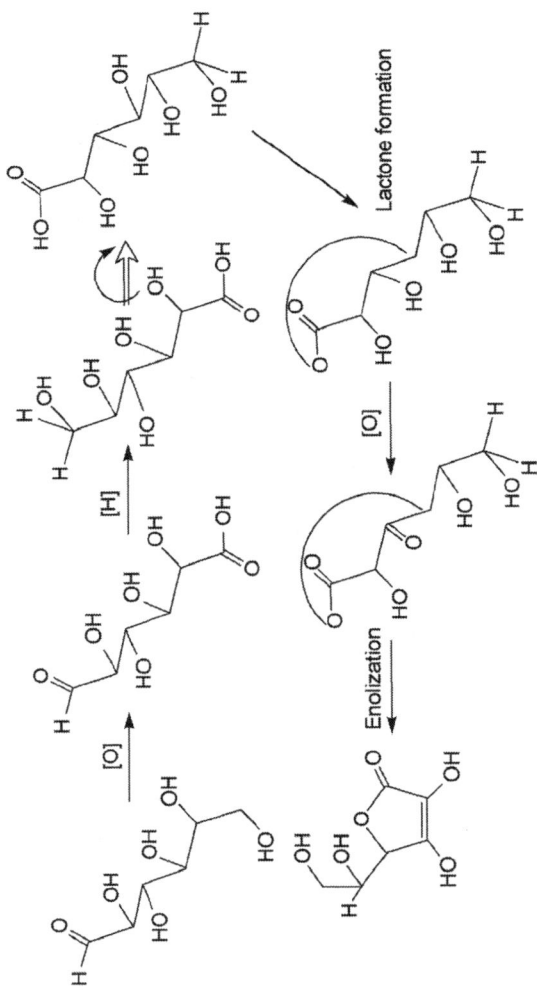

Figure 5.8   Synthesis of L-ascorbic acid

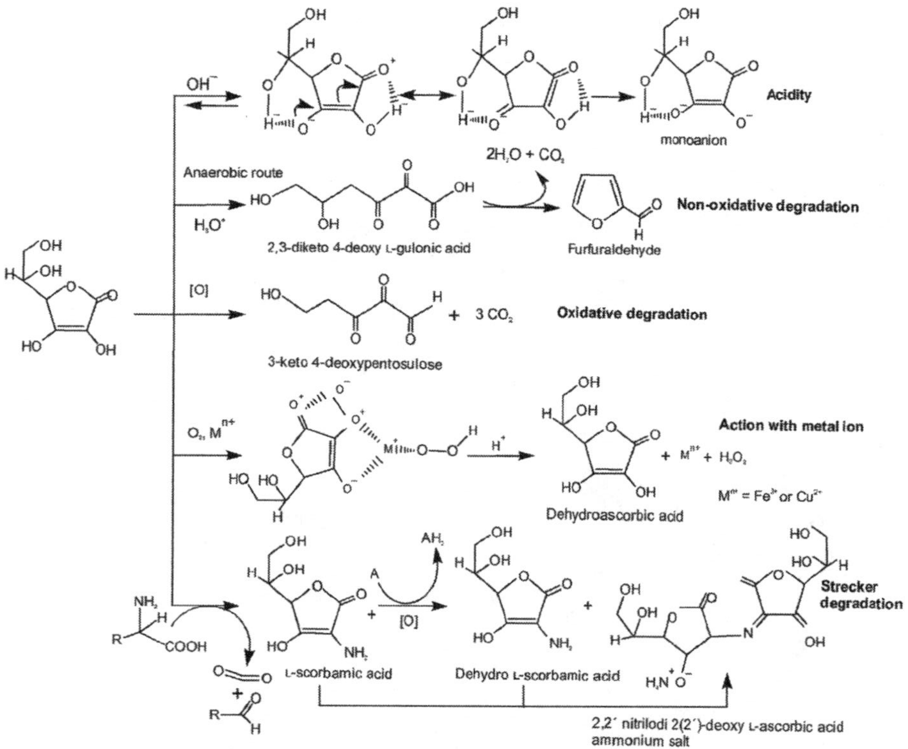

Figure 5.9   Reaction of L-ascorbic acid

BIOLOGICAL FUNCTIONS

❉ It helps the building and maintenance of bone matrix, cartilage, dentine, collagen, connective tissue, etc.

❉ It plays an important role in the removal of iron from ferritin (a protein–iron–phosphorus complex, a form of iron storage) while haemoglobin is formed or red cells are matured.

❉ In the liver, it helps the conversion of folic acid to folinic acid.

❉ It helps the conversion of 3,4-dihydroxyphenylethylamine to noradrenaline.

❉ It cements the ground substances of supportive tissues in the wound healing.

❉ It helps maintaining the resistance to infections.

❉ It promotes growth of body tissues, especially during pregnancy; it helps the growth of foetal and maternal tissues.

❉ It gives protection against free radical damage.

DEFICIENCY DISORDERS

❉ Scurvy—weight loss, weakness, heart palpitations, redness and swelling of gums, loosening of teeth, haemorrhage into the skin and mucous membrane, oedema, hyperirritability, etc.

❉ Metabolism of tyrosine and cholesterol is partially affected.

❉ Absorption and utilization of iron are affected.

RECOMMENDED DAILY ALLOWANCE

National Research Council recommends that children have to take as much as 35 mg. Males and females need 45 mg. Pregnant and lactating women require 60–80 mg.

ANALYSIS OF VITAMINS

Vitamins are analysed applying the spectrocolorimetric method. Each vitamin gives a characteristic absorption at a particular λ_{max} with a specific reagent. The general principle is given in Table 5.2.

Table 5.2   Analysis of vitamins

Vitamin	λ_{max} (nm)	Reagent	Observation (colour)
A	325	Sb_2Cl_3 in $CHCl_3$	Pink
		Cl_3CCOOH in CH_2Cl_2	Blue
D	265	Sb_2Cl_3 in $CHCl_3$	Pink
E	292	Iron (III) salt in 4, 7-diphenyl-1, 10-phenanthrolein	Blue
K	244–249	Sodium diethyl dithiocarbamate in EtONa	Blue
		Ethyl ethanoate with NH_3/KOH	
C	245 (H^+) 265 (Neutral)	2,6-dichlorophenol indophenol	Blue
B_1	235–245	$K_3Fe(CN)_6$ in KOH	Red
B_2	266, 371, 475	$K_3Fe(CN)_6$ in KOH	Red
B_6	325	Indophenol + 2, 6 dichloroquinone + chloroimide	Red

(Contd.)

Table 5.2 (Continued)

Vitamin	λ_{max} (nm)	Reagent	Observation (colour)
Niacin	385	Cyanogen bromide	Red
B_{12}	278	Cyanogen bromide	Red
Folic acid	282	Cyanogen bromide	Red
Pantothenic acid	358	Acidified PhOH + KI + chlorinating agent	Blue
Biotin	234	Acidified PhOH + KI + chlorinating agent	Blue

For the quantitative estimation of each vitamin, the standard solutions with each of the vitamin are prepared. With the appropriate reagent, the intensity values are measured. A standard graph is plotted with intensity values and concentrations. From this plot, the amount of a vitamin in the given food sample is found out.

REVIEW QUESTIONS

Give short answers

1. What are vitamins?

2. Mention the two important characteristics of vitamins.

3. List any four B-complex vitamins.

4. Vitamin A is a _____compound. It is known as _____.

5. List four nutritional sources of vitamin B.

6. Draw the structure of retinol.

7. Night blindness is a deficiency disorder of _____.

8. Vitamin D was discovered by_____ in the year 1919.

9. What is osteomalacia?

10. Mention the two active forms of vitamin E.

11. Tocopherol was identified first by_____and _____.

12. The Nobel laureate Henrik Dam identified_____ first.

13. Name the three forms of vitamin K.

14. In 1912 _____ found out vitamin B as an effective substance to prevent beriberi.

15. Draw the structure of menadione.

Give detailed answers

1. Explain the reactions of vitamin A with reference to

    i.   photooxidation

    ii. photodimerization

    iii. "ene" reaction

2. List the biological functions and deficiency disorders of vitamin A.

3. Write the chemical conversions of provitamins D and D_2 to vitamins D and D_2.

4. Explain the biological functions and deficiency disorders of vitamin D.

5. How does vitamin E react with the following :

    i. $AgNO_3$

    ii. O_2, Fe^{3+}

    iii.   Heat

6. Give the biological functions and deficiency disorders of vitamin E.

7. Write the chemical characteristics of vitamin B_1.

8. Mention the biological functions and deficiency disorders of vitamin B_2.

9. How does vitamin B_2 react with the following:

    i. H_2SO_3

    ii. H_2O, OH^-

    iii.   light

10. Give the biological functions and deficiency disorders of

    i.   Pantothenic acid

    ii.   Folic acid

11. Draw the structure of vitamin B_9.

12. Write the chemical equation for following transformations of vitamin B_6:

    i.   *trans*-amination

    ii.   beta-elimination

    iii.   decarboxylation

13. Write the biological functions and deficiency disorders of vitamin B_{12}.

14. Discuss the following chemical properties of vitamin C:

    i.   Acidity

    ii. Non-oxidative degradation

    iii. Oxidative degradation

    iv. Action with Fe^{2+} or Cu^{2+}

15. List the biological functions and deficiency disorders of vitamin C.

PROBLEMS

1. A sample of food weighing 1.003 g is dissolved in sufficient amount of petether. Then it is made up to 100 mL using chloroform. 2 mL of this solution is analysed to contain 0.0956×10^{-5} g vitamin A. What is the percentage of vitamin A?

    Ans: 4.766×10^{-3}

2. Accurately 1.003 g of a food sample is dissolved in ethanol–water solvent mixture and made up to 100 mL in a standard flask. 1 mL of this solution is taken for analysis and it is found to contain 0.049×10^{-5} g vitamin B. Find out the percentage of vitamin B.

    Ans: 0.04885

3. A fruit sample (1.053 g) is dissolved in ethanol–water solvent mixture and made up to 100 mL in a volumetric flask. The amount of vitamin C in 5 mL of the solution is 0.3633×10^{-3} g. Calculate the percentage of vitamin C.

    Ans: 0.690

4. A fruit piece weighing 1.003 g is dissolved in petether solvent and made up to 100 mL using chloroform solvent. A standard solution of vitamin A is prepared by dissolving 0.048 g vitamin A in 100 mL petether solvent in a standard flask. The standard solutions of different volumes (0.5, 1.0, 1.5, 2.0, 2.5 and 3.0 mL) are taken in different test tubes. Each of these solutions is mixed with 2 mL antimony chloride reagent. The blue layer is used for the measurement of percentage transmittance. The same way, 1 mL of the fruit sample solution is mixed with the reagent and % transmittance is measured. The data are given below. Draw the standard graph and find out the percentage of vitamin A.

Std. soln.	0.5	1.0	1.5	2.0	2.5	3.0	Test solution
% T	77	74	71	67	65	62	74

5. A fruit piece weighing 1.003 g is dissolved in ethanol solvent and made up to 100 mL. A standard solution of vitamin B is prepared by dissolving 0.0511 g vitamin B in 100 mL water in a standard flask. The standard solutions of different volumes (0.5, 1.0, 1.5, 2.0, 2.5 and 3.0 mL) are taken in different test tubes. Each of these solutions is mixed with 2 mL ferric sulphate–KCNS reagent. The red colour layer is taken for the measurement of percentage transmittance. The same way 1 mL of the fruit sample solution is mixed with the reagent and % transmittance is measured. The data are given below. Draw the standard graph and find out the percentage of vitamin B.

Std. soln.	0.5	1.0	1.5	2.0	2.5	3.0	Test solution
% T	92	89	87	83	80	77	81

6

MINERALS AND WATER

INTRODUCTION

Apart from organic nutrients, the body requires inorganic nutrients like water, sodium, potassium, calcium, phosphorus, iron, iodine, etc. that play important roles in metabolic activities. They are builders, activators, regulators, transmitters and controllers. Depending upon the amount, the minerals are classified into three groups.

Group I: Major Minerals

Calcium, magnesium, sodium, potassium, phosphorus, sulphur, chlorine.

Group II: Trace Minerals

Iron, copper, iodine, manganese, cobalt, zinc, molybdenum.

Group III: Trace Minerals (functions, not known)

Fluorine, aluminium, boron, selenium, silicon, cadmium, chromium, vanadium, tin, nickel.

Water accounts for about 60% of the body weight. Water is lost from the body through faeces, urine, sweat and expired air.

WATER, A NUTRIENT

Nutrients are those elements from the food and drinks we consume that perform a sustaining or metabolic function inside our bodies. Essential nutrients must be supplied from an outside source because the body cannot make them in sufficient amounts. Water is an essential nutrient. Without water, human life cannot survive. Water deprivation causes greater hazards than lack of any other nutrient. Such an important substance is seldom considered a nutrient. Many people do not realize the important part water plays in major body functions.

Water: Vital Link to Life

1. Water serves as the body's transportation system. It is the medium by which other nutrients and essential elements are distributed throughout the body. Without this transport of supplies, the body factory would stop functioning. Water also works as the transport for body waste removal.

2. Water is a lubricant. The presence of water in and around body tissues helps defend the body against shock. The brain, eyes and spinal cord are among the body's sensitive structures that depend on a protective water layer.

3. Water is present in the mucous and salivary juices of our digestive systems. This is especially important for moving food through the digestive tract. Persons who experience reduced salivary output soon will realize that foods taste differently and are harder to swallow. As a lubricant, water also is helpful for smooth movement of bone joints.

4. Water participates in the body's biochemical reactions. The digestion of proteins and carbohydrates to usable and absorbable forms depends on the presence of water as part of the chemical reaction.

5.   Water regulates body temperature. Our health and well-being are dependent on maintaining the body temperature within a very narrow range. The human body, which is made of 60–75 per cent water, serves this function quite well. Water itself changes temperature slowly and is able to regulate body temperature by serving as a good heat storage material.

Evaporation of water from the body surface also helps cool the body. Water lost through sweat is barely noticeable and it occurs every day and night. Individuals may lose up to a pint of water each day in this manner. In hot, humid weather or during exercise, increased sweating and loss of water are more visible.

Water Balance

Each day water losses are balanced with water intake. The body has a sophisticated system that works to maintain water balance. Few of us ever experience malfunctioning of this system. Thirst is a trigger that reminds us to take in more water. At the same time our kidneys regulate urinary output.

Is There a Daily Requirement?

Unlike many of the nutrients, there isn't a specific daily recommendation for water intake. Part of the reason is the variability in individuals related to the climate in which they live, physical activity, age, state of health and body size. Under typical circumstances adults may replenish up to six or eight cups of fluid each day.

Typical water output is two-quarts or more of water each day. Water losses in urine account for about three-fourths of daily losses. Remaining losses are in the form of sweat, as tiny water droplets in the air we exhale, and through faeces. Infrequent urination or dark yellow urine may be an indication that we need more fluid intake each day.

Water Sources

Water comes from a variety of sources. All beverages or fluids are a source of water. Even solid foods contain water. Lettuce, celery and other vegetables are composed of 90 per cent or more water. Protein-rich foods such as meat, fish or chicken may contain as much as half to two-thirds their weight in water. Even grain products, which do not seem to be watery at all, may contain up to one-third water.

Fats, such as butter or margarine, and sugar are among the foods that contain the least water. Some water, perhaps one to two cups per day, comes from inside our bodies as a by-product of energy metabolism. This amount is small but significant.

It is important to be aware of fluid intake. Even though solid food is a source of water, additional water from drinking fluids is needed. Besides plain water, juices, milk or other beverages boost fluid intake.

There may be a choice whether the fluid we consume is simply water or an energy-rich beverage that may or may not contain other nutrients. This selection may be based on our need for extra calories/and or additional nutrients.

Special Needs

Under special circumstances, fluid intake and output should be more carefully monitored. Examples of the special circumstances follow:

Infants, young children and older people   Children have lower sweating capacity than adults do. They tolerate high temperature less efficiently. Frequent vomiting and severe diarrhoea in infants and young children quickly can lead to water dehydration.

Older persons may be at increased risk for dehydration because their thirst mechanism may not be as efficient as at younger ages. The influence of medications and the presence of disease are other factors that affect fluid intake and water balance. For both the young and the old, frequent water intake should be encouraged.

Athletes   Of all nutritional concerns for athletes, the most critical is adequate water intake. The athlete's immediate need for water is to control body temperature and to cool working muscles.

Lack of water, above all other nutrients, has the danger to hinder performance and lead to serious complications. For example, fluid loss of two to three per cent of body weight by sweating impairs performance. Fluid losses of seven to ten per cent of body weight may result in heat stroke and death.

Two to three per cent fluid losses in a 150-pound individual represents three to four and a half pounds of body weight or one and a half to two quarts lost water. Marathon runners and other athletes may lose up to three quarts (or six pounds) of sweat per hour.

To prevent dehydration during exercise, athletes should drink fluids before, during and after activity. Even in cold weather, exercise results in sweat production and requires adequate fluid replacement.

Although electrolytes such as sodium are also lost through perspiration, the immediate need is only for water. In most circumstances, sodium and other electrolytes can be replaced after exercise. Seasonings (especially salt) on foods at regular meals usually will do the trick.

What about specially made sports drinks? It is a common belief that the extra electrolytes and sugar in sports drinks provide an edge in competition. Unfortunately some of these special drinks are too concentrated with sugars, electrolytes and flavouring agents that hinder water absorption.

As a general rule, beverages that contain sugar and electrolytes should be diluted with plain water before drinking (Table 6.1).

Table 6.1　Beverages for rapid fluid replacement

Soft drinks	Usage
Water, club soda, Perrier, Seltzer	Use as such
Sports drinks	Dilute 1 part with 2–3 parts water
Juice	Dilute 1 part with 7 parts water
Sodas (diet)	Dilute 1 part with 1 part water
Sodas (regular)	Dilute 1 part with 3 parts water

Source *Sports Nutrition: A Guide for the Professional Working with Active People*. Chicago, IL: American Dietetic Association. 1986.

During exercise or athletic events, thirst is not always a reliable gauge of fluid needs. The best approach is to go into the event with adequate fluid intake.

Six Basic Rules for Fluid Replacement during Sports Events

1. Cooling the water to 40–50°F is best.

2. Plain water is best. Sugar and electrolytes in fluids may slow emptying from the stomach.

3. Don't depend on thirst. Drink ahead of your thirst.

4. Drink water before a sporting event. Two cups of water about two hours before an event is right. Follow this with one cup of water about 15 minutes before the event.

5. Sip water during an event (1/3 to 3/4 cup every 10–20 minutes). The body cannot absorb more than about one cup every 20 minutes.

6. Weigh yourself before and after a sporting event or heavy workout. After the event drink two cups water for every one pound lost.

Water balance in children involved in sports or physical activities is a special concern. Children have lower sweating capacity and less tolerance for hot temperatures. They need frequent fluid intake in order to regulate body temperatures. A water bottle or ready water supply should be kept handy during all sports activities.

Parents, coaches and others should remember that younger children also take longer than adults do to adjust to warm weather following cool winter temperatures. It is especially important to keep a watch on fluid intake during the adjustment time.

Outdoor workers   The combination of hot, humid temperature and physical activity places outdoor workers at risk for water balance. Anyone who works or plays hard outside, especially in hot weather, needs to keep water handy.

Sipping throughout the work period is better than saving up for scheduled meals or breaks. Pay particular attention as you shift from cool weather to warm weather because it takes a few days for your body to adjust to the warmer temperatures.

Water and other elements We often hear recommendations for increasing fibre intake in our daily diets. Adequate fibre intake is helpful for regulating bowel movements and, possibly, lowering serum cholesterol levels.

Persons with low fibre intakes who wish to increase their use of fibre-rich foods should increase fluids as they increase fibre. If not, they can experience extreme discomfort and risk having an impacted bowel.

Dentists recommend fluoridated water for reduction of tooth decay. If community water is not naturally fluoridated, fluoride may be added. Fluoride concentration of one part per million (1 ppm) is considered safe while helping to increase the hardness of tooth and bone tissue. In some areas, natural waters are known to contain too much fluoride, which can result in permanently discolored and mottled teeth.

Water purity in relation to nitrate content, bacterial contamination and other substances is a concern in some areas. Individuals who use private wells or water systems should have their water quality tested on a regular basis.

MINERALS

The minerals that are needed by the body are listed in the Table 6.2 with the details of their sources and biological functions.

The metabolism of some important minerals is given in Figures 6.1, 6.2, 6.3 and 6.4.

Figure 6.1 illustrates the metabolism of calcium. About 30% of dietary calcium is absorbed through the intestine. Vitamin D is necessary for this absorption and it is increased depending on the body need. The absorption is influenced by certain amino acids, lactose sugars and low pH environment. The absorbed calcium is distributed in plasma and in intestinal fluids. The level of calcium in serum is normally 10 mg per 100 mL. This level is maintained strictly by the parathyroid hormone.

The human body has calcium, which forms about 1.5–2.0% of the total body weight. Nearly 99% of the calcium mineral is in the skeletal tissues—bones and teeth—as calcium salts dahllite and apatite. About 1% of the total body calcium is present in the body fluids. It is available in three forms:

i. *Non-diffusible calcium*   It is bound to proteins like plasma proteins, albumins and globulins.

ii. *Diffusible calcium*   It is responsible for the metabolism and function of the bone, nervous system, heart, the membrane system and blood.

iii. *Diffusible calcium complex*   It is in the complexed state with citrate and other substances.

The remaining 70% of the dietary calcium is not absorbed and excreted as faecal calcium. Also, about 20% of the absorbed calcium is excreted as waste matter along with urine.

Table 6.2 Minerals and their biological functions

Minerals	Metabolism	Functions	Deficiency	Source
Calcium	Absorption, aided by Vitamin D. Deposition–mobilization in bone compartments. Absorption and mobilization controlled by parathyroid hormone. Excretion chiefly by faeces.	Formation of bones and teeth Clotting of blood Permeability of capillary walls Contraction of heart muscles Regulation of excitability of nerve cells	Reduced growth Negative calcium balance Bone weakness Parathyroid glands affected Hyperirritability	Milk, vegetables, greens, fish, milk products, etc.
Phosphorus	Absorption along with Ca, aided by Vitamin D. Excretion chiefly by kidney. Parathyroid hormone controlling renal excretion. Balance in blood level. Continuous deposition and reabsorption in the bone formation process.	Formation of bones and teeth Carbohydrate metabolism Constituent of some coenzymes Formation of phospholipid Constituent of nucleic acids	No deficiency is reported	Milk, egg, meat, fish, vegetables, cereals, nuts, pulses
Magnesium	Absorption increased by parathyroid hormone, hindered by excess fat, phosphate, calcium. Excretion regulated by kidney.	Activation of some enzymes Cofactor of enzyme in oxidative phosphorylation	General depression Muscular weakness Other related defects	Wheat, maize, pulses, vegetables, fruits, nuts, meat, fish

Mineral	Absorption/Excretion	Functions	Deficiency symptoms	Sources
Sodium	Absorbed readily. Excreted chiefly by kidney. Controlled by aldosterone, acid–base balance.	Energy source. Supply of minerals to the cell matrix. Regulation of acid–base balance. Regulation of osmotic pressure	General weakness	Table salt, soda, vegetables, fruits.
Potassium	Secreted and reabsorbed in digestive juices. Excretion guarded by kidney according to blood levels.	Regulation of cell content. Regulation of osmotic pressure. Enhancement of relaxation of the heart muscle	General weakness	Vegetables, legumes, fruits.
Chlorine (as chloride)	Absorbed readily. Excreted by kidney.	Acid–base balance. Chloride–bicarbonate shift. Water balance. Gastric HCl–digestion	Hypochloraemic alkalosis. Vomiting, diarrhoea, etc.	Table salt, water.
Sulphur	Absorbed as such and as constituent of sulphur-containing methionine. Excreted by kidney with respect to protein intake and tissue catabolism.	Essential constituent of cell protein. Enzyme activation. High-energy sulphur bonding in energy metabolism		

(Contd.)

Table 6.2 (Continued)

Minerals	Metabolism	Functions	Deficiency	Source
Iron	Absorption, controlled by mucosal block—ferritin mechanism—aided by vitamin C and gastric HCl. Transport and storage of ferritin. Excretion from tissues in minute quantities.	Haemoglobin formation Cellular oxidation to produce ATP	Reduction in growth Pregnancy demands anaemia	Liver, meats, egg, whole grains, vegetables, legumes, nuts
Copper	Bound to an α-globulin as ceruloplasmin. Stored in muscle, bone, liver, heart, kidney and central nervous system.	Haemoglobin synthesis Absorption and transport of iron Maintenance of brain tissues and nervous system	Hypocupremia Nephrosis	Liver, meat, fish, vegetables, cocoa, cereals, nuts, pulses
Iodine	Absorbed as iodides, taken up by thyroid gland with the influence of thyroid stimulating hormone. Excreted by kidney.	Synthesis of thyroxine Regulation of cellular oxidation	Endemic goitre Cretinism	Iodized table salt, seafood
Manganese	Absorption limited. Excretion by intestine.	Activates the reactions like urea formation, protein metabolism, glucose oxidation, lipoprotein clearance, and fatty acid synthesis.	General weakness Inhalation toxicity in the case of miners.	Cereals, grains, soyabean, legumes, nuts, tea, coffee, vegetables, fruits

Mineral		Function	Deficiency	Source
Cobalt	Absorbed as constituent of vitamin B_{12}.	Constituent of vitamin B_{12} Essential factor for the synthesis of red blood cells	Deficiency associated with vitamin B_{12} Pernicious anaemia	Vitamin B_{12}
Zinc	Transported with plasma protein. Stored in liver, muscle, bone and organs. Excreted by intestine.	Forms an important part of some enzymes like carbonic anhydrase, carboxypeptidase, and lactic dehydrogenase Combines with insulin for the hormone storage	Liver-related disease Delayed wound healing Retarded sexual and physical development	Liver, seafood, egg, milk, whole grains
Molybdenum	Minute traces in the body.	Forms a part of the enzyme that converts purine to uric acid Aldehyde oxidation		Meat, milk, leafy vegetables, legumes, grains
Fluorine	Deposited in bones and teeth. Excreted in urine.	Dental health	Dental problems Endemic dental fluorosis	Water, vegetables
Selenium	Active cofactor in cell's oxidative enzymes.	Involved in fat metabolism As "factor 3", along with vitamin E it prevents fatty liver	Liver ailments to some extent	Seafood, meat, grains
Chromium	Improves defective intake of glucose by body tissues.	Glucose metabolism Improves sugar level during fasting	Problems in sugar metabolism Cardiovascular disorder	Meat, whole grains

(Contd.)

Table 6.2 (Continued)

Mineral	Metabolism	Functions	Deficiency	Source
Nickel	Binding by phytate, reduces intestinal absorption.	Forms the protein, nickeloplasmin Associated with thyroid hormone	Cirrhosis Chronic uraemia	Whole-grains, legumes, vegetables, fruits.
Vanadium	Not known clearly.	Helps teeth formation	Lipid metabolism affected	Whole-grains root vegetables, nuts, oils
Silicon	Not known clearly.	Essential agent in formation of bone, cartilage, connective tissues Bone calcification and healing	Bone-related ailments	All vegetables and fruits
Tin	Not known clearly.	Structural element in protein synthesis Associated with cell enzyme system involved in energy metabolism. Tissue growth and wound healing	No specific disorder	Meat, grains, legumes, fruits, acidic juices canned in tin

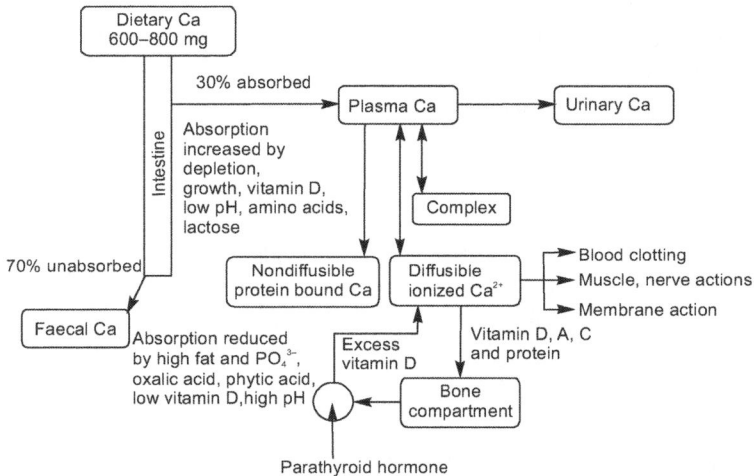

Figure 6.1 Calcium metabolism and distribution of Ca in the body

Figure 6.2 depicts the metabolism of phosphorus. The daily dietary requirement of phosphorus is 900–1200 mg. Almost 70% of the amount is absorbed through intestine along with the calcium. Phosphorus absorption is influenced by excess calcium or other minerals like aluminium and iron that may form insoluble salts. The absorbed phosphorus is distributed in plasma and in intestinal fluids. The phosphorus level in serum is 4.5 mg per 100 mL. The serum phosphorus exists in two anionic forms HPO_4^{2-} (2.1 mEq. per litre) and $H_2PO_4^{-}$ (0.26 mEq. per litre). Always there is a balance between the two anions. Kidney maintains the excretory mechanism for the regulation of serum phosphorus level. Also, the parathyroid hormone plays a vital role in this renal threshold for phosphorus. The action of parathyroid is interdependent with calcium balance. About 90% of the phosphorus absorbed is used for the skeletal tissues (bones and teeth) to form compounds with calcium. The remaining 10% is distributed in the living cells as compounds of proteins, lipids and sugars. These compounds are involved in body building and regulating and energy-producing processes. The unabsorbed 30% of the dietary phosphorus is removed as faecal matter.

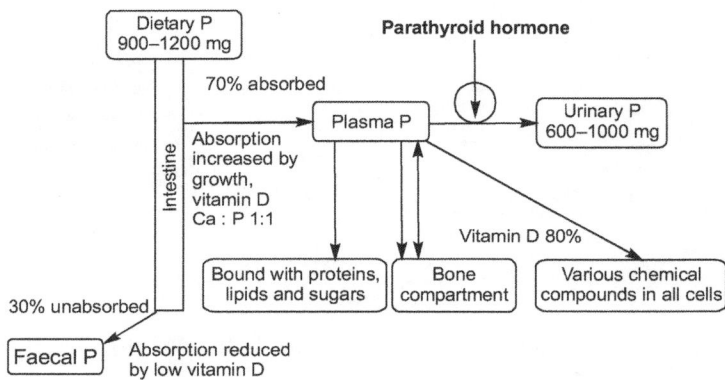

Figure 6.2   Phosphorus metabolism relative distribution and interchange of phosphorus in the body

Figure 6.3 illustrates the metabolism of iron. The dietary requirement of iron per day is 20 mg. It is about 0.005% of the total body weight. Only 30% of the dietary iron is absorbed and the remaining 70% iron is excreted as faecal matter.

The dietary iron is in the ferric-ion form. It is reduced to ferrous ion form by the action of HCl and absorbed. This reduction process is facilitated by vitamin C and a low pH environment. The reduced ferrous ions form a complex with amino acids and carried into the mucosal cells of the intestine. The protein, apoferritin, complexes with ferrous ions to form ferritin that is stored. This comprises 20% of the total body iron. The stored iron in the form of iron–phosphorus–protein complex is transformed into iron–β-globulin complex (transferrin). The transferrin is transported to various cells and

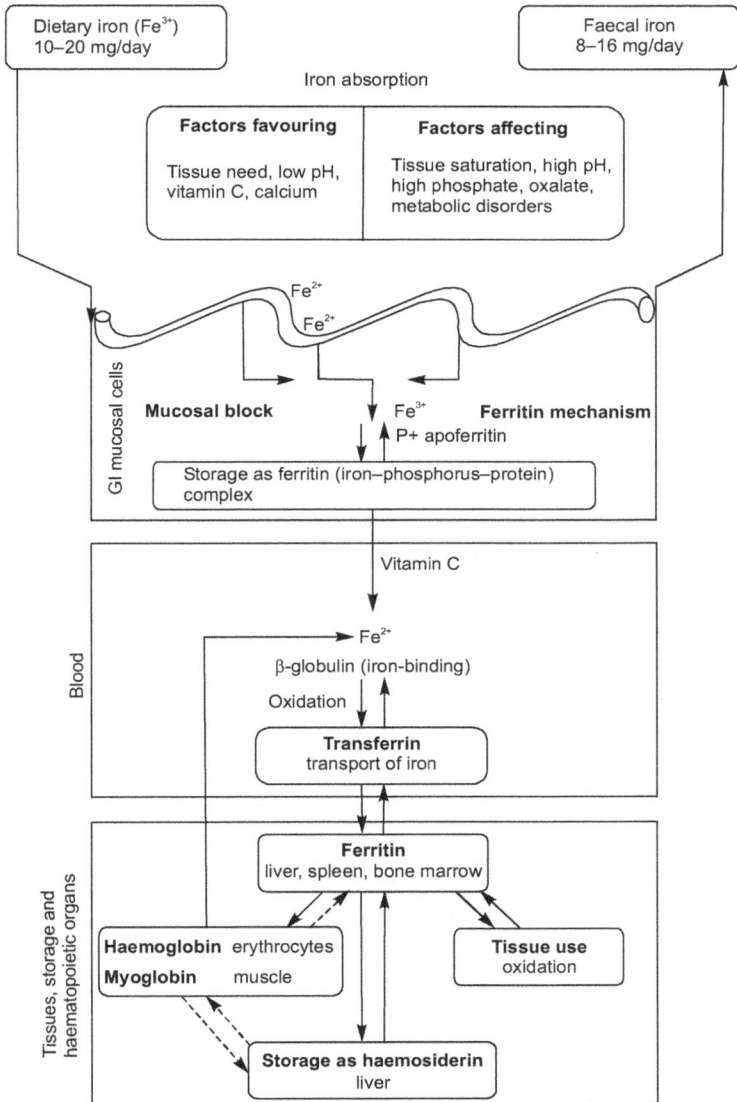

Figure 6.3   Summary of iron metabolism

tissues (in liver, spleen, bone marrow, etc.) and again stored as ferritin. About 1 g of iron that is distributed throughout these cells is in the energy-producing enzyme system. The iron is stored in cells as haemoglobin and myoglobin. The haemoglobins form the essential part of the red blood cells, involved in the oxygen transport to various other cells for respiration and metabolism. Myoglobins also form the part of the muscle cells.

Figure 6.4 sketches the metabolism of iodine. The total iodine in the body is about 50 mg. Approximately 50% of it is present in muscles, 20% in the thyroid glands, 10% in the skin, 6% in the skeleton and 14% in the endocrine tissues, central nervous system and plasma transport system.

Iodine that is actually in the iodide form is absorbed in the small intestine. It is loosely bound to proteins. These iodide-bound proteins are taken to the thyroid gland by blood. About one-third of these proteins are taken up by the help of the thyroid-stimulating hormone (TSH). The remaining two-third of the iodine is excreted through urine.

Iodine in the thyroid gland takes part in the synthesis of thyroxine (a thyroid hormone). The thyroxine is secreted into the bloodstream and is bound to plasma protein for transport in the body cells. The serum level of protein-bound iodine (PBI) is about 8 mg per 100 mL. The thyroxine stimulates cell oxidation by increasing oxygen uptake and enhancing reaction rates of the enzyme system handling glucose. Thus, iodine plays an important role in the overall metabolism of the body.

After cell oxidation, the thyroxine is degraded in the liver and the iodine is excreted in bile as inorganic iodine and gets into the gastrointestinal tract. Thus, iodine is excreted as faecal iodide.

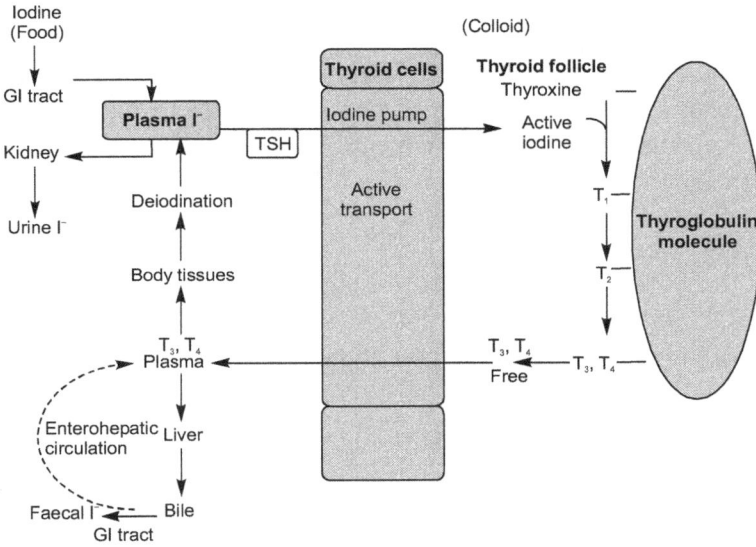

Figure 6.4 Metabolism of iodine, active pump in the thyroid cells and the synthesis of thyroxine

ESTIMATION OF MINERALS

The elements like sodium, potassium, calcium, magnesium, iron, phosphorus, copper, etc. are analysed quantitatively, applying different techniques. The other elements like sodium, potassium and calcium are analysed by flame photometric method. Iron and phosphorus are estimated by spectrophotometric method. Copper and magnesium are estimated by titrimetric method.

Estimation of Sodium

The standard solutions of sodium are prepared in different concentrations. With each solution (known conc.) the flame intensity is measured using a flame photometer. Taking the concentrations of the standard solutions and the corresponding flame photometer readings, a standard graph is plotted. The sample solution is also subjected to this study and the intensity reading is fitted with the standard graph. From the graph, the concentration of the sample solution is determined.

Estimation of Potassium

The method is similar to the one described for sodium. The standard solution is prepared using potassium chloride.

Estimation of Calcium

The estimation is done as described earlier by flame photometry. The standard solution is prepared using calcium sulphate.

Estimation of Iron

The standard solutions of iron are prepared in different concentrations. Each solution (known conc.) is mixed with a known amount of ammonium thiocyanate. The intensities of the resulting red solutions are measured using the spectrophotometer at a particular wavelength. Taking the concentrations of the standard solutions and the corresponding spectrophotometer readings, a standard graph is plotted. The sample solution is also subjected to this study and the intensity reading is fitted with the standard graph. From the graph, the concentration of the sample solution is determined.

Estimation of Phosphorus

The standard solutions of phosphorus are prepared in different concentrations using ammonium phosphate. Each solution (known conc.) is mixed with ammonium molybdate solution.

The resulting yellow colour intensity of each solution is measured using the spectrophotometer at a specific wavelength. Taking the concentrations of the standard solutions and the corresponding spectrophotometer readings, a standard graph is plotted. The sample solution is also subjected to this study and the intensity reading is fitted with the standard graph. From the graph, the concentration of the sample solution is determined.

Estimation of Calcium/Magnesium

A definite amount of the sample solution containing calcium or magnesium ions is titrated with standard EDTA solution at a pH using solochrome indicator. From the titration value, the amount of calcium/magnesium is determined.

Estimation of Copper

The definite amount of the sample solution containing copper ions is titrated with standard thio solution using KI and acid. Starch is the indicator used. From the titration value, the amount of copper is determined.

ESTIMATION OF WATER CONTENT

A known quantity of the food sample is taken in a pre-weighed crucible. It is kept in air-oven for 6–10 hours at 120°C. Then, the crucible is taken out, cooled and weighed. The difference in weight gives the amount of water that is lost during heating. Thus, the percentage of water is secured.

ESTIMATION OF ASH CONTENT

A known amount of the dry, moisture-free food sample is taken in a preweighed crucible. It is heated in a muffle furnace for

15–30 minutes at 300°C. Then, the crucible is taken out, cooled and weighed. The difference in weight gives the amount of chemical substances that are reduced to ash. Thus, the percentage of ash content is found out.

REVIEW QUESTIONS

Give short answers

1. What are the three groups of minerals required for human biological activities?

2. Because of the high specific heat capacity of water it helps _____.

3. List three biological functions of calcium.

4. Mention two nutritional sources of calcium.

5. The metabolic activities of magnesium are _____ and _____.

6. List the deficiency disorders of magnesium.

7. Mention the biological functions of sodium.

8. What is the biological role played by sulphur?

9. What mineral deficiency causes giotre?

10. Mention any two nutritional sources of iodine.

11. Cobalt deficiency leads to _____.

12. List the biological functions of fluorine.

13. Mention two biological functions of molybdenum.

14. Name the disorder that arises due to the deficiency of nickel.

15. What are the biological functions of silicon?

Give detailed answers

1. Write a summary of metabolism of calcium.

2. How is phosphorus metabolized in the body?

3. Discuss iron metabolism.

4. Write a short summary of iodine metabolism.

5. Write the principle of estimation of sodium by flame photometry.

6. How is potassium estimated by flame photometric method?

7. Briefly discuss the method of estimation of iron by spectrophotometry.

8. Discuss the principle of estimation of copper present in food.

9. Give the biological functions of water.

10. How is water content present in food estimated?

PROBLEMS

1. A fresh fruit sample of 12.2822 g is kept in a silica crucible and dried in an air-oven at 200°C for 3–4 hours. The dry sample is found to weigh 2.5320 g. What is the percentage of its water content?

2. A food sample weighing 12.282 g is dried first and then ashed by keeping it in a muffle furnace at 600–800°C for 15–30 minutes. The ash is weighed to be 0.1394 g. Find out the percentage of ash content.

3. A sample of fruit is taken for the analysis of its sodium mineral. A standard solution of sodium is prepared by weighing 0.02953 g NaCl and making up to 100 mL. From this solution other standard solutions of lower concentrations are prepared and taken for flame photometric measurements. Exactly 0.410 g of the ash (4.2768 g fresh fruit) is digested in 100 mL water. 1 mL of the sample solution is taken for the flame photometric measurements. The data are given below. Draw the standard graph and find out the percentage of sodium.

Amount of sodium $\times 10^{-5}$ g	1.161	2.321	3.482	4.642	test solution
Flame photometer reading	58	72	86	100	87

4. A fruit sample is taken for the analysis of potassium by flame photometry. The standard solutions of potassium (four different concentrations) are prepared. Each of these solutions is taken for the flame photometric measurement. Exactly 0.410 g of the ash (4.2768 g fresh fruit) is digested

and dissolved in 100 mL water. 1 mL of the sample solution is taken for the flame photometric measurements. The data are given below. Draw the standard graph and find out the percentage of potassium.

Amount of potassium × 10^{-5} g	1.967	3.934	5.900	7.867	test solution
Flame photometer reading	30	55	75	100	58

5. A food sample (4.2763) is ashed, digested and made up to 200 mL. 10 mL of this sample solution is taken for the calcium analysis by flame photometric method. It contains 1.24×10^{-4} g calcium. Calculate the percentage of calcium present in the food sample.

6. Accurately 10.75 g of a dried fruit is ashed, digested and made up to 100 mL. 10 mL of this solution is taken for the quantitative estimation of phosphorus and it gives 8.1×10^{-3} g P. Find out the percentage of phosphorus.

7

FOOD PROCESSING

INTRODUCTION

Any food item becomes fit for eating only after processing. For example, the fruits like banana, pineapple or papaya need to be peeled off before eating. Rice has to be cleaned (de-weeding, de-husking, steam boiling, etc.) before eating. These preliminary and other stages of operations on food items are collectively called **Food Processing**. There are two main types of processing. One is **Cooking** and the other is **Preserving**. Cooking is done to make the food fit for eating. Preservation is to keep the food free from decay.

COOKING

Most of foodstuff need to be cooked so that the food becomes fit for eating (excepting few fruits, vegetables and nuts). Cooking is a process in which foodstuff are changed desirably in order to make the stuff fit for eating. The changes may be physical or chemical. Peeling off or cutting a fruit will bring about only physical changes whereas roasting or steam boiling of a foodstuff will change food chemically. There are quite a number of processes employed in cooking.

Advantages of cooking

1.   Improves the taste and quality of food

2.   Kills the disease-causing microorganisms

3.   Enhances the digestibility of the food

4.   Makes the food fit for eating

5.   Increases the variety of foodstuff

6.   Concentrates the nutrients by reducing water concentration

7.   Keeps some chemical substances readily available for the body

Cleaning   It is a process in which the food is refined in the following ways:

❋   removal of unwanted portion of the food.

❋   washing with water to remove some dust/toxic/ decayed substances.

There are some disadvantages of this process. Sometimes, valuable nutrients are washed away. For example, rice loses vitamin B while washing.

Peeling and stringing   Removal of the outer skin of a fruit, vegetable or nut. Thus, the non-edible portion or toxic substances are removed. In some cases, the useful nutrients are removed in this process.

Cutting and grating   The foodstuff is cut into pieces so that it becomes handy for eating or cooking. There are different ways of cutting according to the type of food.

Chopping is cutting into no specified shapes.

Mincing is very fine cutting.

Dicing means cutting to uniform cubes.

Slicing is cutting the food into thin pieces.

This process may improve the palatability of the food. But, some loss of nutrients is unavoidable at this stage.

Sieving It is done to remove coarse fibres and insects.

Soaking It is a process in which the food material is completely covered with water or water plus small amount of sodium chloride/carbonate. Water-soluble nutrients may be lost in this process.

Mixing This involves combining two or more food substances homogeneously. But, constituents maintain their identity.

Blending This also involves combining two or more food materials homogeneously but, the constituents lose their identity, and form a single substance.

Beating Mixing two or more foodstuff forcibly such that they combine to give a smooth texture is called beating.

Whipping Mixing two or more food substances forcibly such that they combine to give a smooth texture with frothing (small bubbles) is called whipping.

Mashing This involves crushing the food materials into a smooth, dry and pasty structure.

Stuffing Filling one food substance in the other. For example, stuffing of mashed potatoes in parathas.

Coating Covering a food with a layer of flour, crumbs or other fine powdery substances before cooking.

Blanching   Immersing the foodstuff first in boiling liquid and then in cold water.

Marinating   Soaking the food in a marinade to make it tender or to have fine smell. The marinade may be a liquid prepared for a type of food. For example, meat is marinated with a liquid made up of oil, spices and acid.

Sprouting or germination   This involves soaking the grams in water for few hours so that it starts germinating. At this stage the grams are taken for making salads and other curries.

Fermentation   Transforming the complex substance in the food into simpler ones with the help of beneficial microorganisms. The enzymes of the microorganisms bring about this transformation. For example, curdling, leavening of the flour enhances the edibility of the food.

Cooking Methods

Cooking is one of the processing techniques in order to make the food fit for consumption. Cooking involves wet heating or dry heating or a combination of both. Let us look into the various techniques involved in the methods of cooking of food.

The techniques in wet-heating method are as follows:

1. Boiling
2. Blanching
3. Poaching
4. Pressure cooking
5. Simmering
6. Steaming
7. Stewing

The dry-heating techniques include the following:

1. Baking

2. Frying

3. Grilling

4. Roasting

5. Toasting

6. Sauteing

A combination of any two or more methods is termed braising.

Wet-heating methods

Boiling Cooking the food to boil by immersing in water at 100°C. The foodstuff like rice, egg, potatoes, greens and vegetables, meat, lentils and other cereals are boiled by this method.

It is the simplest method of cooking. Proteins and sugars are denatured sufficiently. Loss of nutrients and flavour, removal of some water-soluble pigments, loss of shape of the foodstuff, use of fuel are the disadvantages.

Blanching Immersing the foodstuff, first in boiling liquid and then in cold water. Foodstuff like potatoes, carrots, etc. are blanched.

Simmering Cooking the food by immersing in water at 80–90°C. While cooking, the vessel is covered with a lid. This is a mild and prolonged boiling method. It is a useful method in which the food is cooked thoroughly and at the same time the stuff does not lose its shape. Loss of some nutrients is the disadvantage of this method.

Poaching Cooking in minimal quantity of water at a temperature of 80–85°C. The temperature is below the boiling point. Egg, fish and fruit are poached generally. It is a quick method of cooking. The foodstuff thus cooked is easily digestible. Loss of water-soluble nutrients cannot be avoided in this process.

Stewing Cooking the foodstuff with a small quantity of water, in a tight-lid pan. It is a type of pressure cooking as steam is generated inside the pan. Initially the heat applied is high. Then, it is brought to simmering temperature. There is no loss of nutrients or flavour. But, it is a time-consuming process.

Steaming Cooking of foodstuff in steam, inside a pan. The food is completely covered by steam while cooking (Figure 7.1). Idly, puttu are normally steam-cooked. The advantage is that there is no loss of nutrients in this cooking. Steam-cooked foods are soft and easily digestible. The cost of the steam pan is the only disadvantage.

Pressure cooking Cooking of foodstuff in steam, inside a pan under a heavy pressure. The pan is so designed that there is pressure formed and thus the food is cooked. The food is completely covered by steam while cooking. The cooking temperature is in the range 130–140°C. The cooking is complete. The advantage is that there is no loss of nutrients in this cooking.

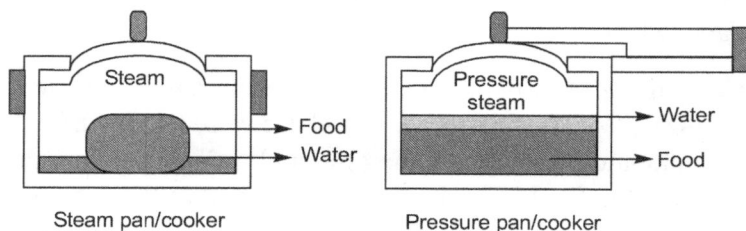

Steam pan/cooker   Pressure pan/cooker

Figure 7.1   Steam and pressure cooking vessels

The cooking time is very short. Fuel consumption is less. Cost of the pressure pan is the only disadvantage. Mishandling of the pressure pan may result in a mishap; one must be aware of the operation procedures.

Dry-heating methods

Grilling or broiling Cooking the foodstuff by direct heating without water. The foodstuff is placed on the grill as shown in Figure 7.2. The food is rotated for uniform cooking and it gets browned. Only some food items like meat, papad, corns and vegetables can be grilled. It is a quick method of heating.

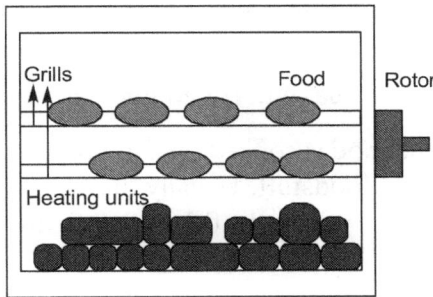

Figure 7.2 Grilling unit

Less heat is required. Flavour of the food gets improved. Charring may be possible if proper care is not taken.

Pan-broiling Cooking the foodstuff on a heated pan that is not covered. For example, chappathi, egg and some nuts are cooked this way. It is a method of fast cooking. It improves the colour and flavour of the foodstuff. Nutrients may be lost by this method. Charring may take place when proper care is not taken.

Baking The food is cooked by the hot air. Basically, it is a dry-heat method. But, steam that comes from the foodstuff is also used (Figure 7.3). For example, bread, cakes and puddings are cooked by this method. The baked foods are brown and crisp on top and soft inside. The flavour and taste of the baked food is always better. Special skills are required to operate the baking apparatus.

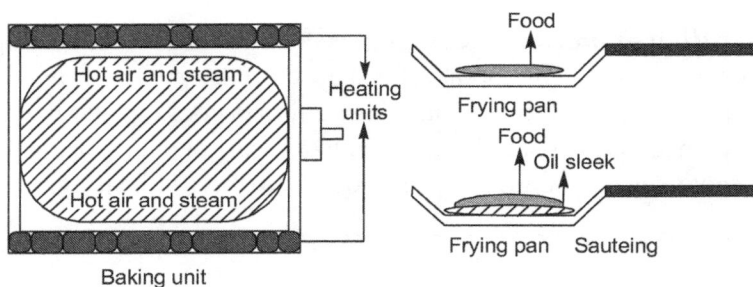

Figure 7.3   Baking, frying and sauteing units

Sauteing   The food is cooked on a hot pan with a sleek of oil at the base of the foodstuff. Usually the food is cooked on all sides by tossing. For example, fish and dosa are cooked by this method.

Deep fat frying   Food is totally immersed in large quantity of oil that is maintained at a high temperature and cooked (Figure 7.4). For example, papads and savouries are cooked by this way. Taste of the food is well improved. Foodstuff is well cooked. Oil consumption is heavy. There are chances for burns and injuries as there is bubbling of hot oil.

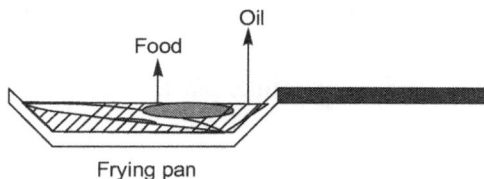

Figure 7.4   Deep fat frying

Microwave cooking In this method the electromagnetic radiation—microwave is used for cooking the foodstuff (Figure 7.5). Energy in the form of radiation is used. Inside the oven, there is source of microwave radiation. It is absorbed by food directly. The water molecules are first excited by the radiation. The water molecules behave as a dipole and begin to vibrate very fast (2500 million times per second). During these vibrations, the molecules experience lot of friction, consequently heat energy is produced. The molecules in the food directly and effectively absorb this heat energy. The efficiency of the microwave cooking depends on the nature of the chemical constituents in food. It is a very fast method of cooking. There is no loss of nutrients. The original colour and flavour of the

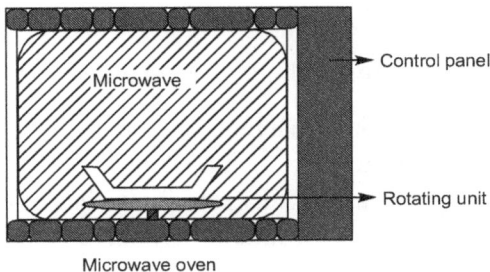

Microwave oven

Figure 7.5   Microwave oven

foodstuff is not changed. Multiple heating of the foodstuff is possible. The cooking unit and specially designed vessels are expensive. It can be used wherever there is electric power supply. The person who cooks must be aware of the operating procedures of this oven.

Effect of Cooking on Nutrients

Carbohydrates The moist starch present in the food gets converted to starch granules, to begin with (Figure 7.6). Later, it gets hydrated to amylose and solubilized into solution when there is water, or gets charred when there is no water. The breakdown products of starch may be easily digested by the biological secretions in the digestive process.

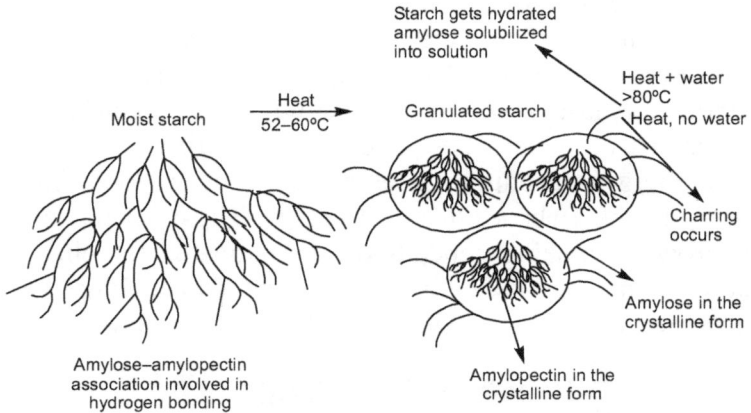

Figure 7.6　Chemical changes in starch on heating

Fat　When fat is heated to a high temperature, toxic polymers are produced (Figure 7.7). They may be hydrocarbons, unsaturated acids, aldehydes, alcohols, lactones, aromatic compounds, etc. These products may form a significant part of flavouring compounds that are formed during the course of the reaction.

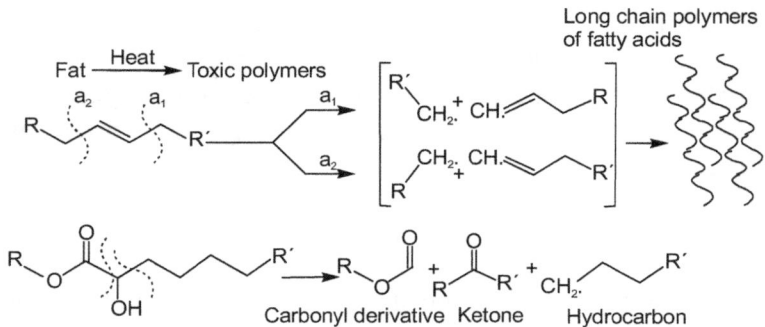

Figure 7.7　Chemical changes in lipids on heating

Protein Proteins get denatured ultimately on heating. It is a gel-formation process to begin with and then denaturation occurs. In gel formation, inter- and intramolecular cross links and sulphur–sulphur cross links are established. But, pH, ionic strength, concentration of proteins, etc. influence this process. Severe heating (like roasting, baking, frying, etc.) may affect the nutritive value of proteins adversely.

$$\text{Protein} \xrightarrow{\text{Heat}} \text{Denatured protein} \xrightarrow{\text{Heat}} \text{Gel-form protein}$$

$$X(\text{Protein})_{\text{natural}} \xrightarrow{\text{denaturation}} X(\text{Protein})_{\text{denatured}}$$
$$\xrightarrow{\text{aggregation}} [(\text{Protein})_{\text{denatured}}]X$$

X moles of protein gets denatured to give X moles of aggregation of denatured protein.

During heating, the end amino group of the essential amino acids (like lysine, arginine, tryptophan, histidine, etc.) reacts with the reducing sugar to form some heterocyclic compounds. This is known as **Maillard Reaction**. This reaction is very unique in the formation of flavouring compounds while food is being processed. Dehydration and/or fragmentation form the flavouring compounds such as furans, pyrones, carbonyls, and acids. Other heterocyclics like pyrroles, pyrazones, oxazoles, thiazoles and sulphur compounds are also formed. The types of reactions are Streckers degradation, condensation, etc. The reaction schemes are given in Figures 7.8.

1.   Formation of pyrrole and pyrazine

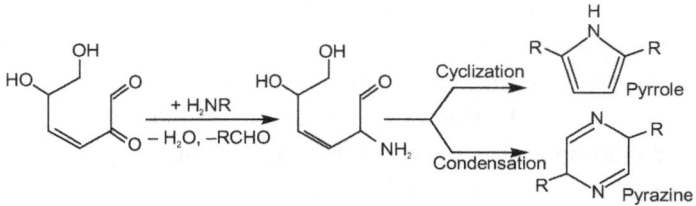

2.   Formation of oxazoles and derivatives

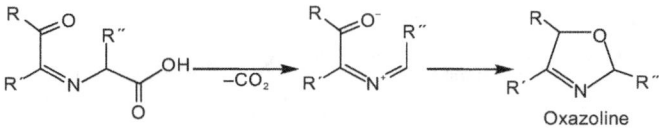

3.   Formation of pyrrolines and pyrrolidines

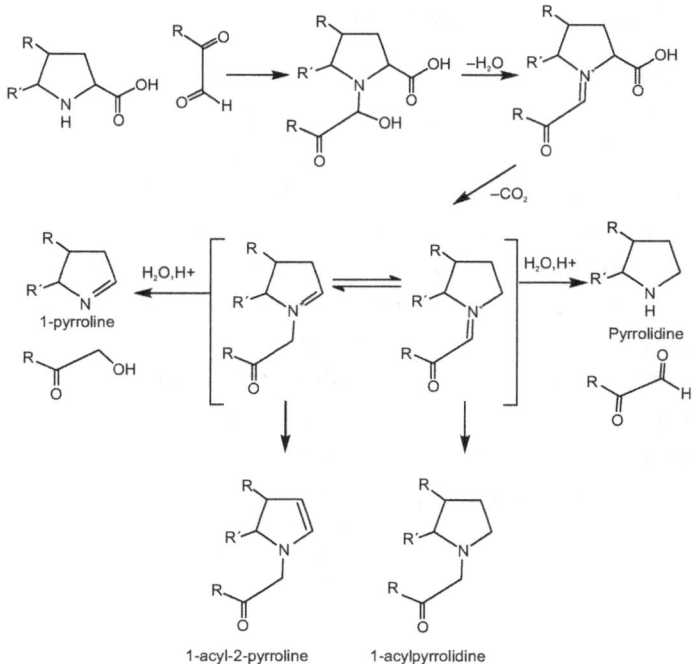

4. Formation of pyrones and reductones:

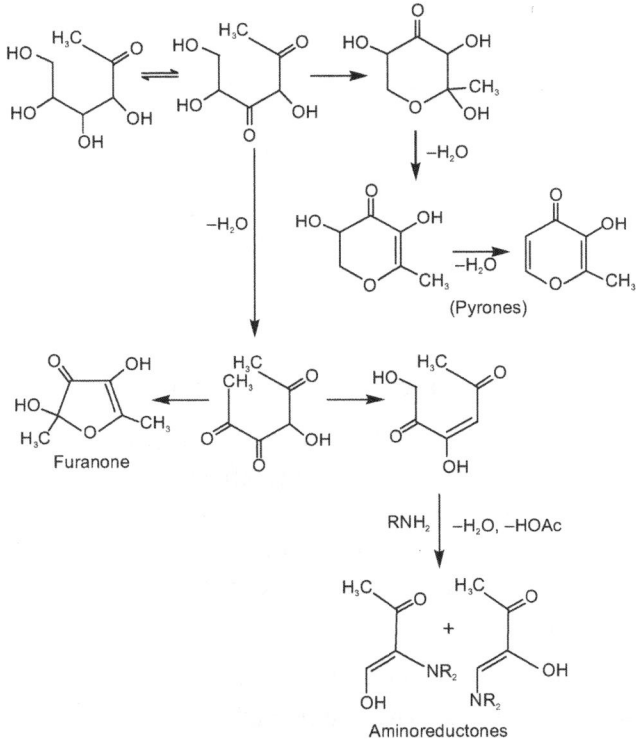

(Pyrones)

Furanone

Aminoreductones

5. Formation of thiazole and thiazoline:

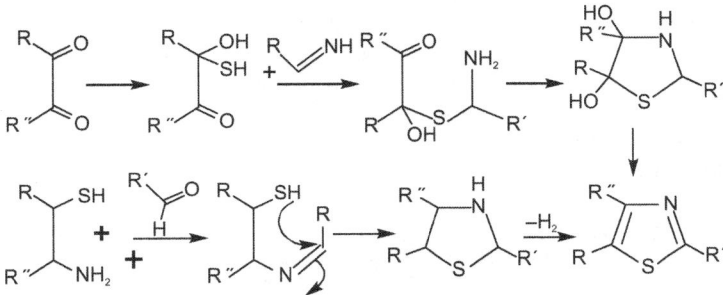

Figure 7.8 Chemical changes in proteins on heating (*Continues*)

6.   Formation of polysulphide heterocyclics:

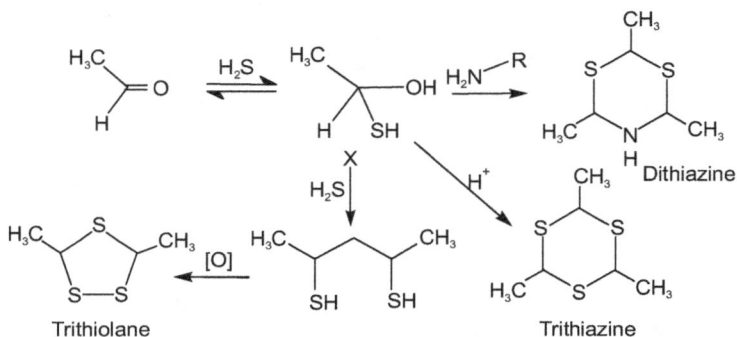

Figure 7.8   Chemical changes in proteins on heating

Vitamins   Water-insoluble vitamins are not affected while cooking in water. But, dry heating affects those vitamins and they are degraded. Water-soluble vitamins get lost while cooking in water. Dry heating deforms those water-soluble vitamins. The chemistry of degradation reactions of some of the vitamins is given here.

Vitamin A   It gets degraded to give many carcinogenic aromatic compounds when subjected to dry heating.

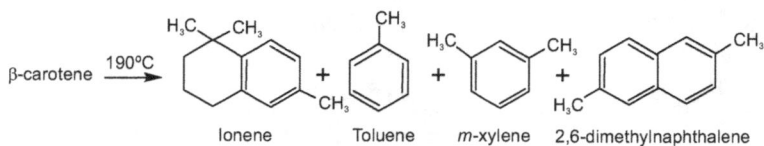

This reaction proceeds by the rearrangement of eight-electron system

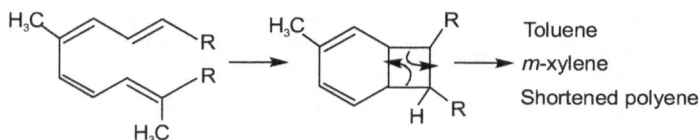

Thiamine (B₁) When it is cooked with baking soda (NaHCO₃) it gets destroyed.

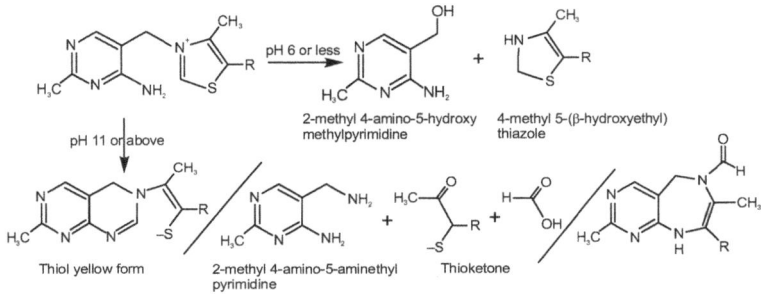

Minerals Minerals like calcium, phosphorus, sodium, potassium, iron, magnesium and other trace elements are lost while cooking for a long time with water. So, care must be taken to avoid such losses during cooking.

Effect of Cooking on Various Foodstuff

Cooking is a type of processing of food. This has great impact on the foodstuff. The nutrients are lost directly or the nutrients are destroyed by some chemical reactions. The following are examples.

Cereals Considerable loss of vitamin B occurs in rice and wheat.

Legumes The growth inhibitors and trypsin inhibitors are destroyed. Nutritive value of the proteins is enhanced.

Oil seeds and nuts Vitamin B content is reduced. Trypsin inhibitors are destroyed.

Milk and milk products Considerable reduction of vitamins A, C and B-complex.

Non-vegetarian food products   Vitamins A and B-complex are lost to a considerable extent when egg is cooked. Fish products will show a considerable reduction in vitamin A. Meat products lose vitamin B-complex while cooking.

Vegetables and fruits   When fresh vegetables are cooked in water, the water-soluble vitamins are reduced in quantity. During dry heating, other nutrients are also destroyed. Fruits may lose the vitamins during dehydration. Canned fruits and fruit juices show an appreciable loss of vitamins C and B-complex.

FOOD SPOILAGE

Any food (whether cooked or fresh) will not remain stable for long. In the course of time, it gets deformed or decayed, and it becomes unfit for eating. This process is called **food spoilage**. It is a natural process by which food undergoes physical and chemical changes to form some harmful products. These changes are effected by microorganisms (like bacteria, fungi, etc.) and macroorganisms (like rodents, insects, etc.).

The spoilage of food by microorganisms cannot be prevented totally. Any food gets spoiled ultimately by the microorganisms. But, spoilage by macroorganisms can be stopped by suitable protection. The spoilage of various foods by microorganisms is given in Table 7.1.

FOOD PRESERVATION

It is a process by which the food is protected from spoilage. Preservation is not a permanent protection for a foodstuff from spoilage. It is only for a short period. The preserved food is in no way superior to the fresh food. The nutrients are not lost from food due to preservation. Food is very important to survival and so it is to be protected from spoilage. Thus, food preservation becomes essential. It is one of the oldest technologies used by human beings. However, the original flavour and the taste may be changed slightly in the preserved food.

Table 7.1   Types of food spoilage by microorganisms

Food	Type of spoilage	Microorganism
Bread	Mouldy	*Rhizopus nigricans, Penicillium, Aspergillus niger*
	Ropy	*Bacillus subtilis*
	Ropy	*Enterobacter aerogenes*
Maple sap and syrup	Yeasty	*Saccharomyces, Zygosaccharomyces*
	Pink	*Micrococcus roseus*
	Mouldy	*Aspergillus, Penicillium*
Fresh fruit and vegetables	Soft rot	*Rhizopus, Erwinia*
	Grey mould rot	*Botrytis*
	Black mould rot	*Aspergillus niger*
	Film yeast	*Rhodotorula*
Pickles	Film yeast	*Rhodotorula*
Fresh meat	Putrefaction	*Alcaligenes, Clostridium, Proteus vulgaris, Pseudomonas fluorescens*
Cured meat	Mouldy	*Aspergillus, Rhizopus, Penicillium*
	Souring	*Pseudomonas, Micrococcus*
	Greening, slime	*Lactobacillus*

(Contd.)

Table 7.1 (Continued)

Food	Type of spoilage	Microorganism
Fish	Discolouration	*Pseudomonas*
	Putrefaction	*Alcaligenes, Flavobacterium*
Egg	Green rot	*Pseudomonas fluorescens*
	Colourless rot	*Pseudomonas, Alcaligenes*
	Black rot	*Proteus*
Chicken flesh	Slime, odour	*Pseudomonas, Alcaligenes*
Concentrated juices	Deflavour	*Lactobacillus, Leuconostoc, Acetobacter*

There are two principles of food preservation:

   i.   Prevention or delay of chemical decomposition

   ii.   Prevention of physical decomposition

Microorganisms, heat, radiation and other external substances may cause the chemical decomposition. This can be prevented by the following procedures:

   ✳   Keeping the food away from microbial infection.

   ✳   Removing the microorganisms from the food.

   ✳   Preventing the growth and the activity of the microorganisms.

   ✳   Destroying the microorganisms.

   ✳   Destroying or deactivating the food enzymes.

   ✳   Preventing the natural oxidation of some of chemical substances by the addition of antioxidants.

Now, we will look at different preservation techniques commonly used today. They are:

❅ Refrigeration and freezing

❅ Canning

❅ Dehydration

❅ Freeze-drying

❅ Salting

❅ Pickling

❅ Pasteurizing

❅ Fermentation

❅ Carbonation

❅ Cheese-making

❅ Irradiation

❅ Chemical preservation

The basic idea behind all forms of food preservation is either to slow down the activity of disease-causing bacteria or to kill the bacteria totally. In some situations, a preservation technique may also destroy enzymes naturally found in a food that cause it to spoil or discolour quickly. An enzyme is a special protein that acts as a catalyst for a chemical reaction, and enzymes are fairly fragile. By increasing the temperature of food to about 150°F or 66°C, enzymes are destroyed.

A food that is sterile contains no bacteria. Unless sterilized and sealed, all food contains bacteria. For example, bacteria naturally living in milk will spoil the milk in two or three hours if the milk is left out on the kitchen counter at room temperature. By putting the milk in the refrigerator you do not eliminate the bacteria already there, but you do slow down the bacterial growth enough that the milk will stay fresh for a week or two.

Preservative methods vary according to the food items and quantity of the items to be preserved. There are household and commercial methods of food preservation available. The principles of food preservation can be broadly classified into two types. They are

    i.   Bactericidal method

    ii.   Bacteriostatic method

Bactericidal method   In this method most of the microorganisms are killed. Examples of bactericidal methods of preservation are cooking, canning, pasteurization, sterilization, irradiation, etc.

Bacteriostatic method   The principle of this method is prevention of multiplication of microorganisms. This may be achieved by removal of water, use of acids, oil, or spices and by keeping at low temperature. The commonly used methods of this principle are drying, freezing, pickling, salting, and smoking. A layer of oil on the top of any food prevents growth of microorganisms such as moulds and yeasts by preventing exposure to air. Spices like pepper and turmeric have some bacteriostatic effect.

Refrigeration and Freezing

Refrigeration and freezing are the most popular food preservation techniques. Refrigeration is done to slowdown bacterial action so that food stays much longer (perhaps a week or two, rather than half a day) in unspoiled state. In the case of freezing, the idea is to stop bacterial action altogether. Frozen bacteria are completely inactive. For example, a bag of frozen vegetables will stay for months.

Refrigeration and freezing are used on almost all foods: meat, fruit, vegetables, beverages, etc. In general, refrigeration has no effect on a food's taste or texture. Freezing has no effect on the taste or texture of most meat and has minimal effect on vegetables. But freezing completely changes fruits (which become mushy). The minimal effects caused by refrigeration account for its wide popularity.

Canning

Since 1825, canning has provided a way for people to store foods for long periods of time. In canning the food is boiled in the can to kill all the bacteria and sealed to prevent any new bacteria from getting in. The food in the can is completely sterile, and so it does not get spoiled. Once the can is open, bacteria enter and begin attacking the food. So, we have to refrigerate the contents after opening (the contents are sterile until you open the container).

We generally think of "cans" as being metal, but any sealable container can serve as a can. Glass jars, for example, can be boiled and sealed. So can foil or plastic pouches and boxes. Milk in a box that can be stored on the shelf is "canned" milk. The milk inside the box is made sterile (using **ultra high temperature (UHT) pasteurization**) and sealed inside the box, so it does not spoil even at room temperature. One problem with canning, and the reason why refrigeration or freezing is preferred to canning, is that the act of boiling food in the can generally changes its taste and texture (as well as its nutritional content).

Dehydration

Many foods are dehydrated to preserve them. Generally, the following are the dehydrated products:

* Powdered milk

* Dehydrated potatoes

* Dried fruits and vegetables

* Dried meat

* Powdered soups and sauces

* Pasta

* Instant rice

Since most bacteria die or become completely inactive when dried, dried foods kept in air-tight containers can last quite a long time. Normally, drying completely alters the taste and texture of the food, but in many cases a completely new food is created that people like these just as much as the original!

Freeze-drying   Freeze-drying is a special form of drying that removes all moisture and tends to have less of an effect on a food's taste than normal dehydration does. In freeze-drying, food is frozen and placed in a strong vacuum. The water in the food then sublimes—that is, it turns straight from ice into vapour. Freeze-drying is most commonly used to make instant coffee powder, but also works extremely well on fruits such as apples.

An Experiment in Freeze-Drying

You probably don't have a good vacuum chamber at home, but you do have a refrigerator. If you don't mind waiting a week, you can experiment with freeze-drying at home using your freezer. For this experiment you will need a tray, preferably one that is perforated. Now you can experiment with apples, potatoes and/or carrots (apples have the advantage that they taste okay in their freeze-dried state). With a knife, cut your apple, potato and/or carrot as thin slices. Then arrange slices on your rack or tray and put them in the freezer. You have to do this fairly quickly or else your potato and/or apple slices will discolour.

The slices should be frozen solid. Over the next week, check the slices. The water in the slices gets converted straight from solid water to water vapour, without going through the liquid state (this is the same thing that mothballs do, going straight from a solid to a gaseous state). After a week or so, the slices are completely dry. To test apple or potato slices for complete drying, take one slice out and let it thaw. It will turn black almost immediately if it is not completely dry.

When all of the slices are completely dry, what you have is freeze-dried apples, potatoes and/or carrots. You can "reconstitute" them by putting the slices in a cup or bowl and adding a little boiling water (or add cold water and microwave). You can eat the apples in their dried state or you can reconstitute them. What you will notice is that the reconstituted vegetables look and taste pretty much like the original! That is why freeze-drying is a popular preservation technique.

Salting   Salting, especially of meat, is an ancient food preservation technique. The salt draws out moisture and creates an environment that is not suitable for bacterial growth. If salted in cold weather (so that the meat does not spoil while the salt has time to take effect), salted meat can last for years.

This technique was the basis for the creation of keg that was used during the voyage time of Columbus. Salting was used to preserve meat up through the middle of this century, and was eventually replaced by refrigeration and freezing. Today, salting is still used to create salt-cured meat and dry fish. They are fried in cooking oil and used as side dishes.

Pickling   Pickling was widely used to preserve meat, fruits and vegetables in the past, but today it is used almost exclusively to produce "pickles," or pickled cucumbers. Pickling uses the preservative qualities of salt combined with the preservative qualities of acid, such as acetic acid (vinegar). Acid environments inhibit bacteria. To make pickles, cucumbers are soaked in 10-per cent salt water (brine) for several days, then rinsed and stored in vinegar to preserve them for years.

Pasteurizing   Pasteurization is a compromise. If food is boiled, all bacteria are destroyed and the food becomes sterile. But, this significantly affects the taste and nutritional value of the food. When the food (almost always as a liquid is pasteurized only certain bacteria are killed (but not all) and certain enzymes are disabled. Thus the taste is not changed fully. Commonly pasteurized foods include milk, ice cream, fruit juices, beer and non-carbonated beverages. Milk, for example, can be pasteurized by heating to 62.8°C for half an hour or 72.8°C for 15 seconds.

Ultra high temperature (UHT) pasteurization completely sterilizes the product. It is used to create "boxes of milk" that you see on the shelf at the grocery store. In UHT pasteurization, the temperature of the milk is raised to about 141°C (285°F) for one or two seconds, sterilizing the milk.

Fermenting Fermentation uses yeast to produce alcohol. Alcohol is a good preservative because it kills bacteria. When you ferment grape juice you create wine, which will last quite a long time (decades if necessary) without refrigeration. Normal grape juice would mould in days.

Carbonating Carbonated water is water in which carbon dioxide gas has been dissolved under pressure. By eliminating oxygen, carbonated water inhibits bacterial growth. Carbonated beverages (soft drinks) therefore contain a natural preservative.

Cheese-making Cheese is a way of preserving milk for long periods of time. In the process, the milk in cheese becomes something completely unlike milk, but cheese has its delicious properties. Cheese-making is a long and involved process that makes use of bacteria, enzymes and naturally formed acids to solidify milk proteins and fat and preserve them. Once turned into cheese, milk can be stored for months. The main preservatives that give cheese its longevity are salt and acids. The basic steps in cheese making are as follows:

　※　First, milk is inoculated with lactic acid bacteria and rennet. The lactic acid bacteria convert the sugar in milk (lactose) to lactic acid. The rennet contains enzymes that modify proteins in milk. Specifically, rennet contains rennin, an enzyme that converts a common protein in milk called caseinogen into casein, which does not dissolve in water. The casein precipitates out as a gel-like substance that we see it as curd. The casein gel also captures most of the fat and calcium from the milk. So the lactic acid and the rennet cause the milk to curdle, separating into curds (the milk solids, fats, proteins, etc.) and whey (mostly water). About 3.5 kg of milk may yield only about 500 g of cheese—the weight that is lost due to the removal of water from milk.

❊   The curd and whey are allowed to soak until the lactic acid bacteria create a lactic acid concentration that is just right. At that point, the whey is drained off and salt is added.

❊   Now the curds are pressed in a cheese press—lightly at first to allow the escape of the remaining whey, then severely (up to a ton of pressure) to solidify the cheese.

❊   Finally, the cheese is allowed to age (ripen) for several months in a cool place to improve its taste and consistency. A sharp cheddar cheese has been aged a year or more. During this time, enzymes and bacteria continue to modify proteins, fats and sugars in the cheese. As you can see, cheese-making is complicated. It produces a product that preserves milk proteins and sugars with acids and salt.

Note   The holes in Swiss cheese occur during ripening. Swiss cheese is ripened in a cool place for several weeks, then put in a warm place 21°C or (70°F) for four to six weeks, where special bacteria ferment the remaining lactose and produce carbon dioxide bubbles in the cheese.

Irradiation   Nuclear radiation is able to kill bacteria without significantly changing the food that contains the bacteria. So if the food is sealed in plastic containers and then irradiated, then the food becomes sterile and can be stored on a shelf without refrigeration. Unlike canning, however, the taste or the texture of the food does not change significantly on irradiation. Irradiation of meat could prevent many forms of food poisoning. However, many people are apprehensive about "nuclear radiation." Therefore, irradiated food is not very common in our country.

Chemical preservatives The use of chemical substances is the commonly used and also the most abused method of food preservation. Chemical preservatives serve as anti-microbial, antioxidant or both. They also minimize the damage to some essential amino acids and the loss of some vitamins. Anti-microbials prevent the growth of moulds, yeast, and bacteria whereas antioxidants prevent foods from browning, or developing black spots and becoming rancid. Food that is rancid when consumed, may not make one, fall sick, but it makes the food smell and taste bad. The common chemical preservatives used are sulphites, nitrites, calcium propionate, disodium EDTA, BHA, BHT, citric acid, etc. These chemicals if used in permitted dosage/quantity, may not be harmful. The other method is use of ozone. The use of passive air ozonation has reduced bacteria counts up to 300%.

Sulphites They are used primarily as antioxidants to prevent or reduce discoloration of light-coloured fruits and vegetables, such as dried apples and dehydrated potatoes. This chemical is used to preserve food items such as baked goods, beer, wine, tea, condiments, vinegar, dairy product, processed seafood products, grain products, gravies, sauces, jam, jellies, nut products, plant protein isolates, dried fruit, fruit juices, fruits, vegetables, filled crackers, soup mixes, sugar, sweet sauces, etc. In some countries, the use of sulphites is banned on fruits and vegetables intended to be eaten raw, such as in salad bars and also in thiamine-rich food such as enriched flour. Sulphite-sensitive people are prone to develop breathing difficulty, stomach ache, and occasionally anaphylactic shock.

Nitrites   The nitrites are a favourite preservative of meat processors even though its excess use is restricted in many countries. This chemical stabilizes the red colour of the meat and gives the meat a fresh appearance. Nitrites are used as a preservative for meat and meat products. Nitrite salts can react with certain amines in food to produce nitrosamines, which are known to cause cancer. Addition of sodium ascorbate or sodium erythorbate inhibits nitrosamine formation and reduces the problem of nitrosamines. The use of nitrite and nitrate has decreased greatly because of refrigeration and restriction on the amounts used. Even though nitrite and nitrate cause only a small risk, it is always better to have fresh meat and meat product.

Butylated hydroxyanisole (BHA)   It is a phenol and antioxidant, used as a preservative in cereals, chewing gum, butter, meat, baked goods, beer, snack foods, potato chips, vegetable oil, etc. It slows down the development of off-flavours, odours, and colour changes caused by oxidation. Even though it is considered as safe, some studies showed that it produces cancer in experimental animals.

Butylated hydroxytoluene (BHT)   It is also a phenol compound similar to BHA, and has similar action. BHT retards rancidity in oils and is used in cereals, chewing gum, potato chips, and oils. It is better to avoid this or substitute it with less harmful methods because of its potential carcinogenicity.

Sodium propionate   This preservative is used in bread rolls, pies, and cakes. It prevents growth of microorganisms.

Citric acid   It is one of the popular, cheap and safe antioxidants used. It is used in the preservation of fruit juices, cheese, margarine and salad dressings.

EDTA   It is chemically called ethylenediamine tetraacetic acid. It is an agent, which helps to separate metallic ions. It removes metal impurities in food and thus prevents rancidity. This safe chemical is used in preventing rancidity in salad dressing, margarine, sandwich spreads, mayonnaise, processed fruits and vegetables, canned shellfish, and soft drinks.

Heptyl paraben   This is used in beer and non-carbonated soft drinks as a preservative. This preservative is safe but rarely used.

Lecithin   It is a natural substance present in animal and plant tissues used mainly as emulsifier, but it is also used to retard rancidity in baked goods, margarine, chocolate, and ice creams.

Phosphoric acid   This acid is formed on exposure of phosphorus to air. It is used in baked goods, cheese, powdered foods, cured meat, breakfast cereals, and potatoes. Excessive consumption of phosphates could lead to osteoporosis.

Propyl gallate   It is an antioxidant. It is used for the preservation of vegetable oil, meat products, potato sticks, chicken soup base, and chewing gum. It causes cancer in experimental animals. It shall preferably be avoided.

Sodium benzoate   This is a granular salt used as preservative in fruit juices, carbonated drinks, jellies, margarine, fast-food burgers, and pickles. It is a time-tested preservative used for centuries.

The use of preservatives effects the following changes.

* S–S disulphide bonds are reduced

* carbonyl compounds are formed

* ketone groups are reacted

* respiratory mechanism of microorganisms are inhibited

REVIEW QUESTIONS

Give short answers

1.   What is food processing?

2.   Define cooking.

3.   Briefly explain the term preservation.

4.   Explain briefly the following terms:

   i.   Mashing

   ii.   Blanching

5.   Cooking involves the two heating methods, _____ and _____.

6.   Identify the wet-heating methods in the following:

   i.   Simmering

   ii.   Blanching

   iii.   Grilling

   iv.   Baking

   v.   Stewing

   vi.   Sauteing

7.   What is Maillard reaction?

8.   Draw the structures of the following:

   i.   Pyrrole

   ii.   Pyrazine

   iii.   Oxazoline

9. Explain the chemical conversion of methanal to dithiazine.

10. What happens to vitamin A when it is heated?

11. Define food spoilage.

12. Mention one microorganism that is responsible for the spoilage of each of the following food items:

    i.   Bread

    ii. Pickles

    iii.   Fish

13. Mention the two principles of food spoilage.

14. Write the chemical names of the two preservatives that are used to preserve food.

15. Lipid undergoes degradation on heating to give _____ and _____.

Give detailed answers

1. Explain the term "cooking". List five advantages of cooking of food.

2. Briefly explain the following terms:

    i.   Mashing

    ii. Cutting and grating

    iii. Marinating

    iv. Fermentation blanching

3. Discuss the wet-heating methods.

4. Explain the process of microwave cooking.

5. Discuss the chemistry of starch on heating.

6. How are lipids chemically changed on heating?

7. Outline the formation of the following compounds by Maillard reaction:

    i.   Pyrrole

    ii. Pyrrolidine

    iii. Furanone

    iv. Thiazine

8. How is thiamine chemically modified on heating?

9. Explain the effect of cooking on various foodstuff.

10. Discuss the following methods of food preservation:

    i.   dehydration

    ii. freezing

    iii. radiation

    iv. use of preservatives

8

FOOD ADDITIVES

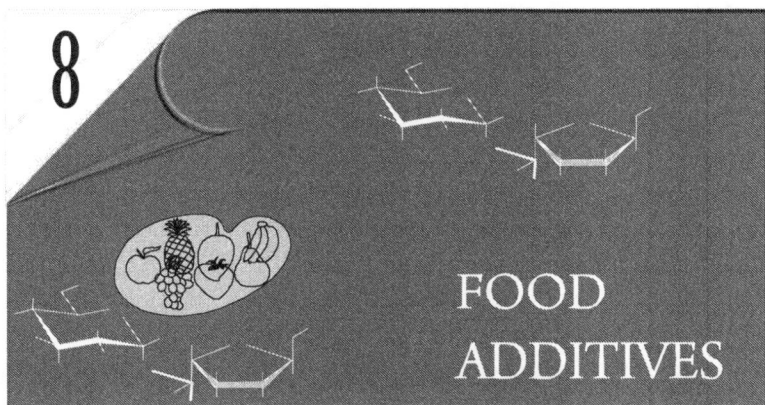

INTRODUCTION

The chemical substances that are added to food from external source are termed **food additives**. The additives are used mostly for commercial reasons—to make the foodstuff sweet or attractive, to impart colour and to have longer shelf life without any chemical damage, etc. Thus, the foodstuff gains market value. The types of additives are sweeteners, flavourmatics, colourants, antioxidants, emulsifiers, acidulants, etc.

THE CHEMISTRY OF SWEETENERS

They are sweet-imparting substances. Table sugar is a sweet-carbohydrate called sucrose. It is made up of two smaller sugar molecules (glucose and fructose) joined and so it is also known as a disaccharide. Here, we consider the sweeteners that are but artificial.

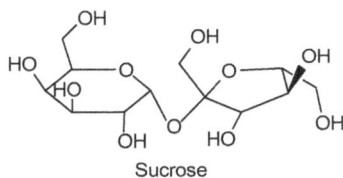

Sucrose

Notice the two six-ring hexagons joined in sucrose. Similar six-ring structures are seen in many artificial sweeteners also.

Intense Sweeteners

These have a diverse range of chemical structures and are much sweeter than sugar and other bulk sweeteners such as xylitol. Intense sweeteners are generally used in products to reduce calories or for making tooth-friendly sweets. Blends of intense sweeteners are increasingly being used in foods. Combinations of sweeteners can be chosen to provide a taste profile far superior to that of a single sweetener. This concept results in the use of smaller amount of sweeteners (Figure 8.1).

* Aspartame is a dipeptide. It is a compound of two amino acids joined (aspartic acid and phenylalanine). Only very small amounts of it are needed for sweetening purposes. It is 200 times sweeter than sucrose (sugar).

* Acesulfame-K is a synthetic compound that acts as a synergist with other sweeteners. This means that a mixture of two sweeteners can be sweeter than an equal quantity of either sweetener alone. It is 200 times sweeter than sucrose (sugar).

* Saccharin is a synthetic compound. The taste of saccharin is not ideal when used alone or at high concentrations. It synergizes well with acesulfame-K and is often used in this way. It is 300 times sweeter than sucrose (sugar).

* Cyclamate is not as sweet as most of the other intense sweeteners and its taste is not ideal at higher concentrations, but it is well-suited for blending with other sweeteners. It is 30 times sweeter than sucrose (sugar).

✳ Sucralose is modified sucrose. In the body it is not used for energy and so whilst it retains the sweet taste of sugar it has no calories. It is also one of the strongest of the artificial sweeteners. It is 600 times sweeter than table sugar.

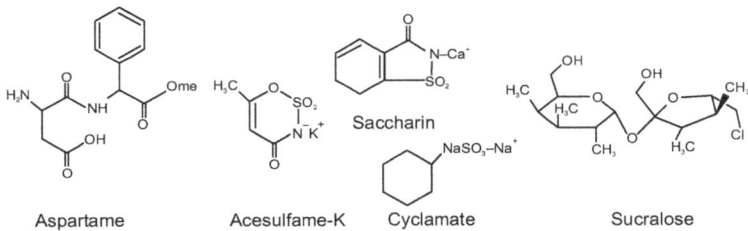

Figure 8.1   Structures of some intense sweeteners

Bulk Sweeteners

This group of sweeteners is predominantly composed of the polyols. They are derivatives of normal sugars and exhibit carbohydrate-like structure (Figure 8.2) and functionality. The polyols can often be used as direct replacements for sugar. Many of the polyols occur naturally, but most are produced commercially by hydrogenation of their corresponding sugar precursors. Polyols are suitable for diabetics by virtue of their reduced glycaemic index. They are also reported to play a role in actively reducing the risk of tooth decay. These are 60% as sweet as table sugar. They may act as a laxative if consumed in excess and a warning to this effect is printed on package labels.

Erythritol is hydrogenated erythrose. It has a high negative heat of solution, but its cooling effect is limited by its moderate solubility. It is 60% as sweet as table sugar.

Isomalt is hydrogenated isomaltulose. It has a relatively low solubility. It is 45% as sweet as sugar (sucrose).

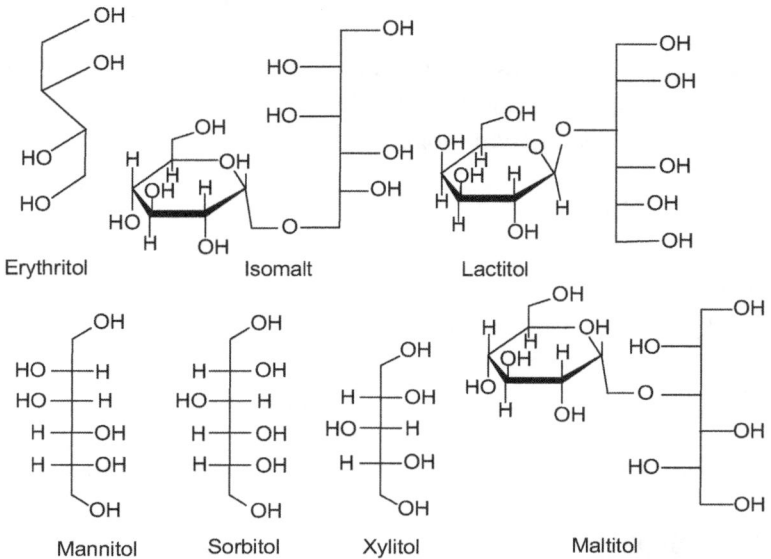

Figure 8.2   Structures of some bulk sweeteners

Lactitol is hydrogenated lactose. It has a high solubility. Due to a low negative heat of solution, it also produces one of the lowest cooling effects of the polyols. It is 35% as sweet as sucrose.

Maltitol is hydrogenated maltose. It has a high solubility and a moderate cooling effect. It is 80% as sweet as sugar (sucrose).

Mannitol is hydrogenated glucose. It has a very low solubility and a low cooling effect. It is 50% as sweet as sugar (sucrose).

Sorbitol is hydrogenated glucose. It has a high solubility and reasonably high cooling effect. It is 60% as sweet as sugar (sucrose).

Xylitol is hydrogenated xylose (wood sugar). It is the sweetest of the polyols with sweetness equal to that of sugar.

It also has a high solubility and provides the greatest cooling effect of all of the polyols. It is 100% as sweet as sugar (sucrose).

CHEMISTRY OF FOOD COLOURS

The colour of a food that is found in nature is due to some characteristic chemical substances. Normally, the colour-imparting substances have many conjugated double bonds. At these positions the electronic transitions like $n-\pi^*$, $\pi-\pi^*$ and $\sigma-\sigma^*$ occur. The colour that is shown by a substance is due to the absorption of its complementary colour. There are three basic colours (blue, green, and red). They mix in different proportions to give three complementary colours (magenta, cyan, and yellow). Many foods have natural colours. Artificial colours are due to the colouring substances (colourants) that are added from external sources. The colourants have no nutritive value, but are used to make the food appear more attractive. For example, ice creams are made in many attractive colours. Food colours are divided into three main types.

Natural Colours

These are obtained from natural sources such as grasses, leafy vegetables, fruit skins, roots and seeds of plants. Animals can also be a source of food colourings. Cochineal, or carminic acid, is a red colour that is obtained from the bodies of certain scale insects. These feed on cactus leaves and their bodies are commercially harvested in Africa, Spain and Central America. Their bodies are dried and crushed to extract the red colouring.

Nature Colours

Obtaining colours from natural sources can be costly and their quality can vary. To overcome this, chemists have found ways to make identical colours in the laboratory. This improves their purity and may also cost less. Nature-identical colours are exactly the same molecules found in natural sources but they are made synthetically. The main chemical classes are:

* ❋ flavonoids, found in many flowers, fruits and vegetables

* ❋ indigoid, found in beetroot

* ❋ carotenoids, found in carrots, tomatoes, oranges and most plants. Carrots contain an orange molecule called beta-carotene, which is part of this group.

Most natural and nature-identical colours can dissolve in oil but do not dissolve in water. This means it is difficult to add them directly to foods. They are usually processed to form their sodium or potassium salt. This makes them soluble in water and suitable for use in foods. They may also be dissolved in oil and incorporated into water-soluble beadlets.

Synthetic Colours

They do not occur in nature and have been made in a factory. They have been carefully tested to make sure that they are safe. The main examples of synthetic colours are:

* ❋ azo dyes, such as amaranth (colour for black currant jams).

* ❋ other dyes, such as, quinoline (quinoline yellow), xanthene (erythrosine), triarylmethanes, indigoid, (indigo carmine).

Synthetic colours are usually water-soluble and can be used in foods without any further processing.

How much Colouring should be in Food?

The law in Europe allows 43 colours with E-numbers to be used in foods. It also lists the foods that may be coloured and the maximum levels of colour that may be added to those foods. The variety of colours allows many different shades to be obtained. Some colouring substances are shown in Figure 8.3.

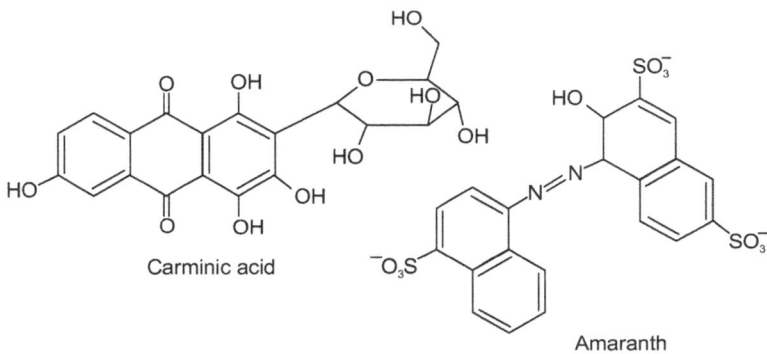

Carminic acid

Amaranth

Figure 8.3   Some colouring substances

Synthetic colours are much brighter than natural colours and so are needed in only very low concentrations. Typically just 10–50 mg in a kilogram of food.

Natural colours are less intense than synthetic colours and so need to be used in higher concentrations. Natural colours are used in concentrations in the range of 0.05–10 g (10–10,000 mg) per kilogramme of food.

Nature-identical colours vary in usage levels but can be very efficient. For example, beta-carotene is used at levels of 1–30 mg per kg of food (Figure 8.4). Synthetic colours are used as they are more stable than natural colours. These synthetic colours are suited to foods and are more stable until they are consumed.

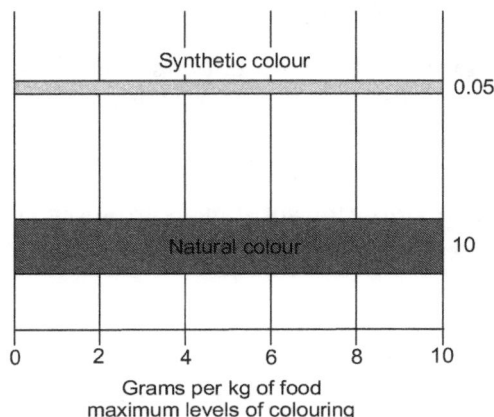

Figure 8.4   Permitted levels of colourants

Beta-carotene is used in making orange squash. Just imagine how much of beta-carotene is required to prepare 150,000 litres of orange squash? That is 1 litre every day for over 400 years? It is amazing to note this great demand of colourants.

Safety

The safety of food colourings is a controversial subject. Concern has been expressed, in particular, about synthetic colours. However, there is no logical or scientific reason to suggest that the coloured chemicals present in nature are any safer than those, which are manufactured. All additives, including those used to colour food, have to be tested and shown to be safe. Nevertheless, some people are sensitive to certain colours and there are claims of links with hyperactivity, asthma and other allergic reactions. Colours that are allowed to be used in foods are strictly tested. Some common food colourings are shown in Table 8.1.

Table 8.1 List of permitted colourants

Name	Description	Food
Curcumin	Orange-yellow colour that is extracted from the roots of the turmeric plant.	Curry, fats and oils, processed cheese, etc.
Riboflavin	Also known as vitamin B_2. It can be obtained by fermenting yeast or synthesized artificially. In foods, it is used as an orange-yellow colour.	Sauces, processed cheese and foods with added vitamins such as bread.
Tartrazine	Yellow coloured synthetic azo dye. This colouring sparks controversy as some groups suggest it causes behavioural problems in children.	Is no longer widely used. Now rarely used in curries and some ready-meals.
Beta-carotene	Orange-yellow colour found in plants such as carrots, tomatoes and oranges.	Soft drinks, margarine, butter, yoghurt.
Plain caramel	Dark brown to black colour. The most common colouring. 90% of all colouring used is caramel. Obtained by the heating of sugars.	Cola drinks, confectionery, baked-foods, ice cream, chocolate, beers, vinegar and whisky.
Amaranth	Dark purple synthetic colour, similar to black currants.	Powdered soup, jam, ice cream, instant gravy.

FLAVOURING AGENTS

The acceptability of any food product greatly depends on the impression of taste when it is eaten. Our sense of taste is really a combination of two of our senses, taste and smell. Both of these senses are responses to certain chemicals. How do we taste? Taste is a complex mixture of flavours and aroma, or smell.

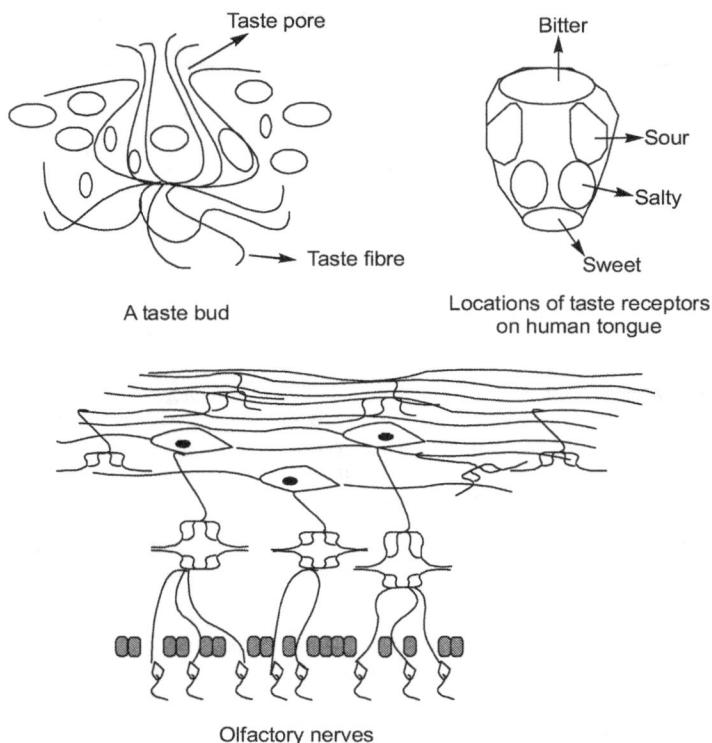

Figure 8.5 The body sensors of taste and smell

The receptors for the human sense of taste are located on the tongue and on the soft palate (Figure 8.5). There are just five stimuli to which these receptors respond. These are sweet (as in sugar), sour (as in acidic substances like lemon juice), bitter (strong coffee or quinine in tonic water), salt (table salt) and umami (monosodium glutamate, savouries, soya sauce, crisps).The active sites of the receptors have specific size to which the chemical substances of appropriate geometry fit-in. Thus, the taste is realized.

Our sense of smell makes up a big part of how well we taste food. Flavour molecules in the food enter the air in the nose and are detected by millions of receptors that feed information to the brain. The active sites of the receptors have specific size to which the chemical substances of appropriate geometry fit-in. Thus, the smell is realized. Chewing helps to transfer more odours from the mouth to the back of the nose. The area, which is sensitive to smell, is located at the back of the nose where several million receptor cells per square centimetre respond to thousands of chemicals in the food.

Sight plays an unexpectedly important role in our perception of flavours. The taste of a colourless, shapeless food is extremely difficult to recognize. We may need visual "clues" to enable us to identify taste and flavour accurately. The brain interprets signals from taste, smell and even vision before turning them into an impression of the food's taste. Different people will find different tastes pleasant or unpleasant. Flavourings are added to food products to give, enhance or intensify flavour.

Some of the flavouring agents, permitted limits and their m.p./b.p. are listed in the Table 8.2.

Table 8.2   List of flavouring substances

Flavouring chemicals	Flavour	Permitted limit (ppm)	m.p. /b.p. (°C)
Vanillin	Vanilla	31.5	81.5 m.p.
Ethyl vanillin	Vanilla	16.6	76.5 m.p.
Citral	Lemon	17.6	229
Benzaldehyde	Cherry/almond	84.8	180
Cinnamic aldehyde	Cinnamon	110.7	252
Methyl salicylate	Root beer	129.7	221
Amyl acetate	Banana	78.4	142

(Contd.)

Table 8.2 (Continued)

Flavouring chemicals	Flavour	Permitted limit (ppm)	m.p. /b.p. (°C)
Amyl butyrate	Banana	23.7	185
Cinnamic aldehyde	Cinnamon	110.7	252
Methyl salicylate	Root beer	129.7	221
Amyl acetate	Banana	78.4	142
Amyl butyrate	Banana	23.7	185
Ethyl butyrate	Strawberry	31.1	120
Safrol	Root beer	16.9	236
Menthol	Mint	111.2	215
Aldehyde-C18	Coconut	17.6	263
Diacetyl	Butter	17.3	88
Ethyl oenantnate	Grape	8.1	187
Allyl caproate	Pineapple	11.7	187
Aldehyde-C14	Peach	8.1	297
Carvone	Spice, mint	190.3	230
Linalool	Spice	30.2	199
Citronallal	Rose	114.2	206
Methyl phenyl acetate	Honey	2.4	222
Aldehyde-C8, C9	Citrus		
Cyclohexyl butyrate	Pineapple	8.3	212
α-Furfuryl mercaptan	Coffee		
Linalyl acetate	Citrus	5.4	220

ANTIOXIDANTS

When food is exposed to air, the oxygen present in air combines with the chemical constituents of food. That is how the oxidation reaction occurs. It is a destructive process. Thus, the food may lose its nutritional value and the chemical composition of the food is also changed.

The oils and fats become rancid on oxidation. A cut apple becoming brown in colour is due to the oxidation of the sugar molecules. The chemical substances that are added to such easily oxidizable food in order to prevent the oxidation are called antioxidants.

Chemistry of Antioxidants

Fats and oils or food containing them are likely to get oxidized in the presence of oxygen and light to form peroxides. These peroxides impart foul smell to the oil or fat. This phenomenon is called rancidity. It is prevented by the addition of antioxidants.

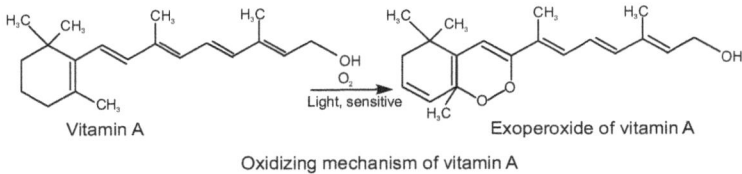

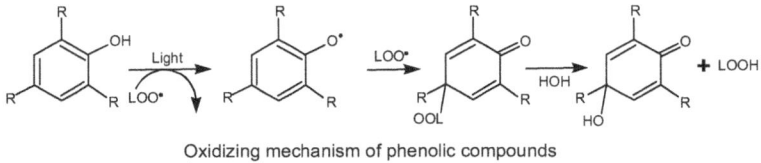

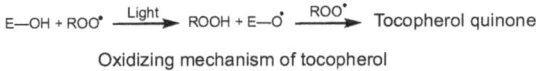

Oxidizing mechanism of vitamin A

Oxidizing mechanism of phenolic compounds

$$E{-}OH + ROO^{\bullet} \xrightarrow{\text{Light}} ROOH + E{-}O^{\bullet} \xrightarrow{ROO^{\bullet}} \text{Tocopherol quinone}$$

Oxidizing mechanism of tocopherol

These antioxidants undergo oxidation with oxygen more readily than the active substances present in foods. The following chemical substances are used as antioxidants: vitamin A, vitamin E and some phenolic compounds. The mechanism of oxidation of these antioxidants are given here.

The number of antioxidants are very limited in number. The chemical behaviour of natural and synthetic antioxidants are the same. So, they are used in combination. Table 8.3 lists some typical antioxidants and their uses.

Table 8.3   Some antioxidants and their uses

Antioxidants	Typical food	How it works
Ascorbic acid	Beers, cut fruits, jams, dried potatoes	Helps prevent the food from going brown by preventing oxidation.
Tocoperols	Oils, meat pies obtained from soya beans and maize	Reduces oxidation of fatty acids and some vitamins
Butylated hydroxyanisole (BHA)	Oils, margarine, cheese, crisps	Helps prevent the reaction that breaks down fats and causes the food to go rancid.
Citric acid	Jam, tinned fruits, biscuits, alcoholic drinks, cheese, dried soup, lemons	Helps reduce the reactions that can discolour fruits. Used to regulate pH in jams and jellies

EMULSIFIERS

The oil and water do mix homogeneously under normal conditions. But, they mix homogeneously in presence of a substance called emulsifier. In some foods, emulsifiers are necessary in order to keep the water portion and the oil portion mixed thoroughly. Emulsifiers can help make the food appealing. Emulsifiers have better impact on the structure and texture of food. Emulsifiers may help prevent the growth of moulds in some foods.

Foodstuff Containing Emulsifiers

The following are the foodstuff containing emulsifiers:

Biscuits, toffees, bread, extruded snacks, chewing gums, margarine, breakfast cereals, coffee whiteners, cakes, ice-creams, topping powders, dried potatoes, peanut butter, soft drinks, chocolate coatings, caramels, etc.

Types of Emulsions

There are two types of emulsions.

  i.   oil-in-water

  ii. water-in-oil.

The first type has the oil droplets dispersed in water medium and in the second type the water droplets are dispersed in oil medium.

The emulsifier molecule has two distinct ends. One is the hydrophilic end which binds with polar end of any other molecule that constitutes the dispersion medium or dispersed phase whereas the other end is the lipophilic end which binds with the nonpolar end of any other molecule that forms part of either dispersion medium or dispersed phase. Thus, the dispersion of oil-in-water or water-in-oil is smooth and homogeneous. The emulsifier mechanism is illustrated in the Figure 8.6.

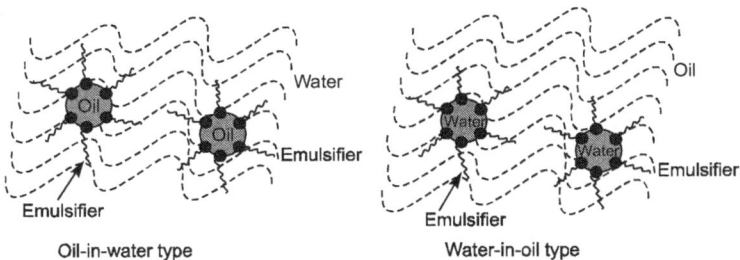

Oil-in-water type   Water-in-oil type

Figure 8.6   Mechanism of emulsification

The emulsifier molecule will form a coat at the surface of the dispersed phase using its polar or nonpolar end depending on the nature of the dispersed phase. This molecular arrangement is called micelle. Milk is a natural emulsion of oil-in-water type. In this the fat droplets are dispersed in water medium homogeneously with the protein emulsifier.

Manufacture of Emulsifiers

The most common types of emulsifiers are monoglycerides which are produced by the reaction between fatty acids and glycerol. The other types of emulsifiers are esters of lactic acid and they are produced by esterification of lactic acid with mono- or diglycerides. Most widely used commercial emulsifiers are lecithin, mono- and diglycerides of fatty acids. The chemistry of manufacture of some of the emulsifiers is discussed here.

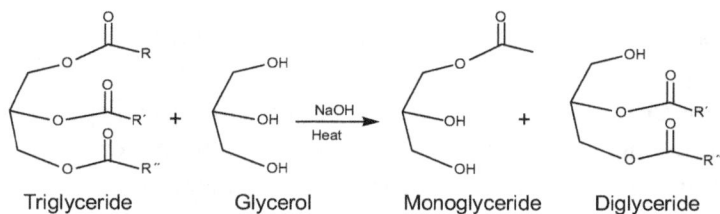

Triglyceride   Glycerol   Monoglyceride   Diglyceride

The triglyceride on reacting with glycerol in presence of sodium hydroxide, mono- and diglycerides are formed. These two products are separated by vacuum distillation. These mono- and diglycerides are used as emulsifiers.

The monoglyceride reacts with succinic anhydride in the presence of NaOH at 80°C to yield succinylated monoglyceride which is used as emulsifier in baked products. Also, the ethoxymonoglyceride obtained by the reaction of monoglyceride with epoxide and NaOH at 150°C and 80 psi is used in the baked product.

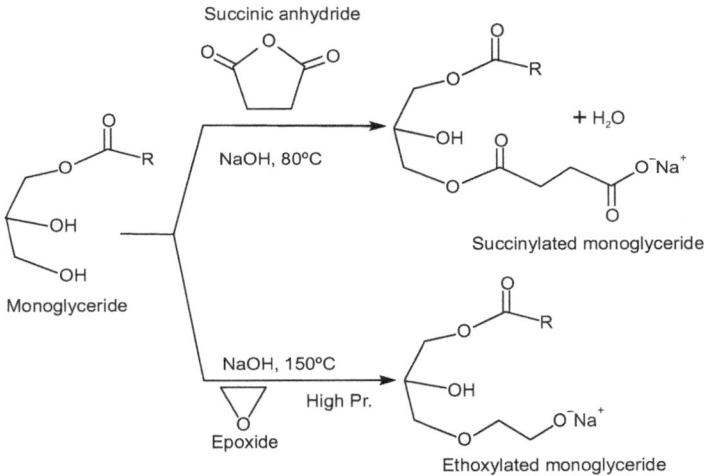

The emulsifier of ester derivative of alcohol can be prepared by the reaction of stearic acid with sorbitol in acidic medium at 225–250°C. This type of emulsifier is used in confectioneries. The polysorbate obtained by the reaction of sorbitan monostearate is also used as an emulsifier in cake manufacture. The reaction is shown on page 210.

Functions of Emulsifiers in Food

Emulsifiers stabilize the emulsion by minimizing the aggregation of fat globules. The softness of the baked foods is remarkably improved by the use of emulsifiers. Wheat dough gains strength by the interaction of emulsifier molecules with gluten. The consistency of fat-based products is improved by controlling the crystallization of fat.

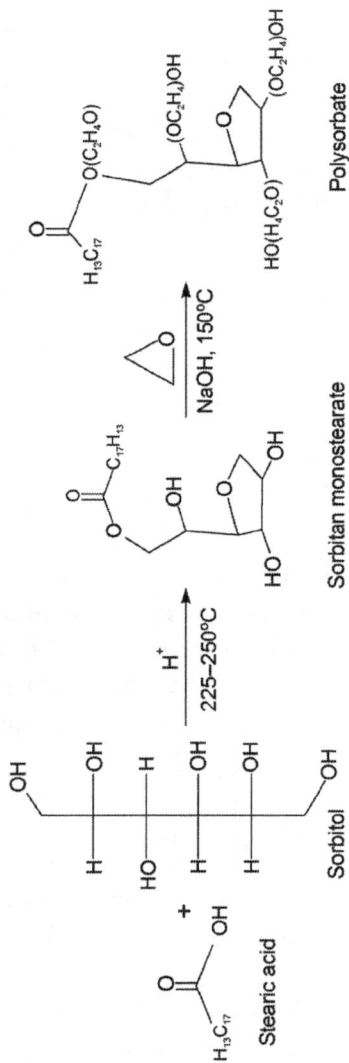

Sorbitol + Stearic acid → (H⁺, 225–250°C) → Sorbitan monostearate → (NaOH, 150°C) → Polysorbate

ACIDULANTS

The acids that are added to foods to improve the taste and to increase the shelf life are called acidulants. The foods that contain acids naturally are very sharp in taste. For example, orange, lemon, apple and tomato are some of the foods containing natural acids. The presence of the acids impart characteristic sour taste to these foods.

Acids, alkalis and buffers play a big role in food industry. Acids have been used for centuries as flavouring and antimicrobial agents. For example, the citric acid is used in soft drinks and pharmaceuticals. Vinegar is used in pickles and other confectionery products. Let us briefly discuss the chemistry of some acidulants.

Acetic acid

It is the main constituent of vinegar. In Latin *acetum* means vinegar. It is a corrosive, pungent-smelling liquid with a boiling point 118°C. It is miscible in water, ethanol, ether. Its acidity is characterized by the pK_a value of 4.78. It is used in pickles and confectionery items. It has excellent bacteriostatic property and so it is used as a preservative. It is used for separating casein from milk by coagulation process. Acetic acid is manufactured by fermenting the liquor of 10–15% alcohol using the bacteria *Mycoderma aceti* in air. The 3–7% solution of acetic acid is called vinegar.

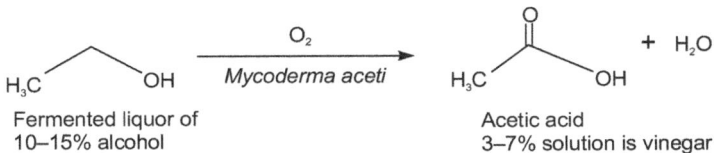

$$H_3C-OH \xrightarrow[\textit{Mycoderma aceti}]{O_2} H_3C-C(=O)-OH + H_2O$$

Fermented liquor of
10–15% alcohol

Acetic acid
3–7% solution is vinegar

Citric Acid

It is widely used in food industries. It is a colourless crystal with a melting point 153°C. It is soluble in water, ethanol extracted from lemon and lime fruits. Commercially it is prepared by fermenting molasses using the mould *Aspergillus niger* or *Citromyces prefferianus*.

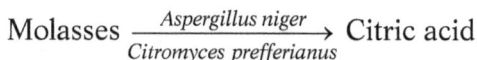

$$\text{Molasses} \xrightarrow[\text{Citromyces prefferianus}]{\text{Aspergillus niger}} \text{Citric acid}$$

Citric acid is formed in nature by the reaction of acetyl-CoA with oxaloacetic acid. It is extracted from lemon or any other citrus fruits.

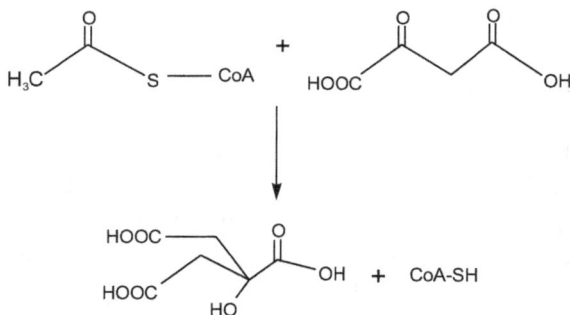

It adds taste to the soft drinks. It generates optimum conditions for the formation of gels, jellies, confectionery, etc. It imparts stability to the emulsions like cheese and dairy products. It prevents the browning of salads. It acts as an antioxidant in oils and fats. It helps in preserving the meat and modify the texture while processing.

Lactic Acid

It is a colourless syrupy liquid with a boiling point of 122°C (m.p. 18°C). It is sour in taste. It is prepared by fermentation process of souring of milk using the microorganism *Lactobacillus bulgaricus* at 45°C in presence of chalk. Calcium lactate is formed. It is then treated with dilute sulphuric acid to give lactic acid. It is widely used in the preparation of boiled sweets, pickled foods and baked foods.

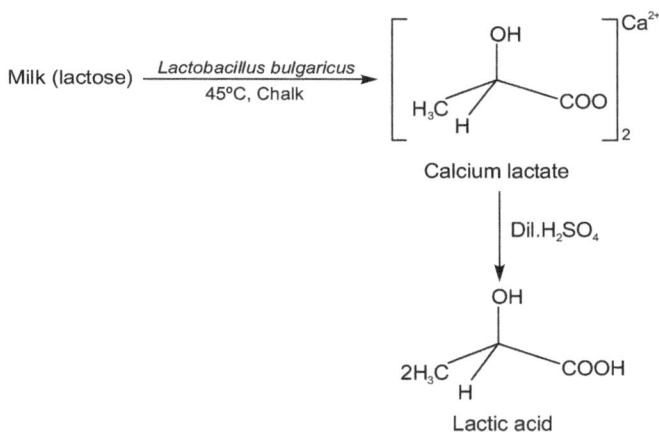

Calcium lactate

Lactic acid

Malic Acid

It is found naturally in apple, pears, tomatoes, bananas and cherries (Latin *malum*=apple). As early as 1785 Schule prepared this compound from apple. The modern method is by the hydration reaction of maleic acid using dilute sulphuric acid.

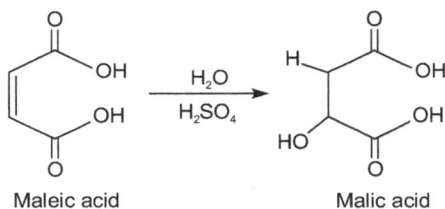

Maleic acid   Malic acid

This acid is used in preparing the low-calorie drinks. It is more expensive than citric acid.

Phosphoric Acid

It is the second most widely used acidulant. It is manufactured from phosphate rock (mined in North Africa and North America). It is a tribasic acid with acidity values ($K_1 = 7.5 \times 10^{-3}$, $K_2 = 6.2 \times 10^{-8}$, $K_3 = 1 \times 10^{-12}$). It forms the main chemical constituent of the soft drink, Coca Cola. It has a harsh, biting taste. It is used as buffers, for baking items and emulsifying salts in the preparation of cheese.

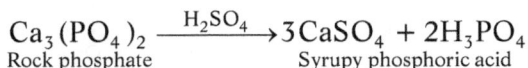

$$\underset{\text{Rock phosphate}}{Ca_3(PO_4)_2} \xrightarrow{H_2SO_4} 3CaSO_4 + \underset{\text{Syrupy phosphoric acid}}{2H_3PO_4}$$

Tartaric Acid

It is found as the potassium salt of juice of grape, tamarind and berry. "Argol" that separates as crust during the fermentation of grape juice, is the impure salt, potassium tartrate. This salt is boiled with chalk. The calcium tartrate thus formed is treated with dilute sulphuric acid. Tartaric acid is formed as colourless crystals (m.p. 167°C). It is soluble in water and alcohol. It is optically active. Tartaric acid is used in the manufacture of emulsifiers. The salt of tartaric acid, potassium hydrogen tartrate is used as acidulant in baking and in confectionery.

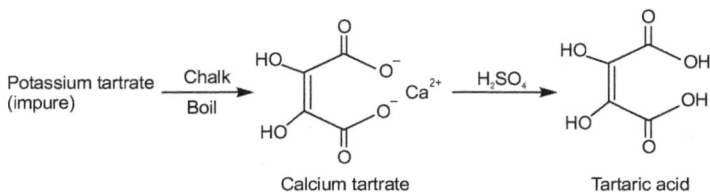

Potassium tartrate (impure) — Chalk/Boil → Calcium tartrate — H_2SO_4 → Tartaric acid

REVIEW QUESTIONS

Give short answers

1. What are sweeteners? Give two examples of artificial sweeteners.

2. Define:

    i. intense sweeteners

    ii. bulk sweeteners.

    Give one example each.

3. Isomalt is _____. It is_____ as sweet as sucrose.

4. Hydrogenated glucose is termed _____. It is_____ as sweet as table sugar.

5. What are synthetic colourants? Give two examples each.

6. Natural colorants are used in the range of_____ per kilogram of food.

7. Define flavouring agents.

8. The characteristic smell of banana is due to the flavourmatics_____ and _____.

9. Name the flavouring substances that are responsible for the smell of

    i. coffee

    ii. grape

10. What are antioxidants? Name two food items to which antioxidants are added.

11. Name any three antioxidants.

12. What are emulsifiers? Mention their types.

13. Name any three food items that commonly contain emulsifiers.

14. What do you mean by acidulants? Mention any three acidulants.

15. Give the acidulants that are added to each of the following food items:

    i.   apple

    ii.   soft drink

    iii.   cheesecake

Give detailed answers

1. Define intense sweeteners. Draw the chemical structures of the following:

    i.   cyclamate

    ii.   sucralose

    iii.   acesulfame-k

    iv.   saccharin

2. What are bulk sweeteners? Draw the structures of the following:

    i.   mannitol

    ii.   maltilol

    iii.   erythritol

3. What are colourants? Draw the structures of

    i.   carminic acid

    ii.   amaranth

4. Write a short account of natural and synthetic colourants.

5. Enumerate the colour and the food items to which the following colourants are used:

   i.   curcumin

   ii. riboflavin

   iii. tetrazine

   iv. plain caramel

6. Draw the location of taste receptors in tongue.

7. Write briefly the mechanism of flavouring action.

8. How do antioxidants work in food? Just list the uses of the following antioxidants:

   i.   ascorbic acid

   ii. tocopherol

   iii. citric acid

   iv. butylated hydroxyanisole

9. Write a short note on emulsifiers. Discuss briefly the uses of the following acidulants:

   i.   tartaric acid

   ii. citric acid

10. Brief the principle of manufacture of lactic acid.

FOOD ADULTERATION AND TESTING

INTRODUCTION

Food Adulteration refers to the intentional addition of some unwanted substances to food or to the removal of valuable substances from food. This adulteration has become a menace and has posed a lot of problems to the humankind. This practice is silently killing the healthy society. People suffer from many ailments. This practice is the handwork of antisocial elements that try to cheat people. The present-day consumer goods, right from the foodstuff up to the medicines that are commercially available, are adulterated in some form or other. We, the consumers are to be aware of this fact.

This chapter aims at making people aware of the varied aspect of food adulteration—legal aspects regarding food adulteration and prevention, the common food adulterants, the techniques to analyse the adulterants, etc.—so that the consumers are less likely to be cheated.

LEGAL ASPECTS OF FOOD ADULTERATION AND PREVENTION

To safeguard the people from the health hazards posed by the practice of adulteration, it is necessary to put forth strict check and control over the quality of foods. International and national standards are available for checking the quality of various foods.

International Codex Alimentarius Commision is the principal organ of a worldwide food standard programme. It is a joint venture of FAO and WHO. The general principles of the Codex Alimentarius are:

❊   To collect the internationally adopted food standards.

❊   To protect consumer health.

❊   To guide people about the food requirements.

❊   To give provision to food hygiene, food additives, contaminants, labelling and methods of analyses.

In India there are different standards for checking the quality of food and other consumer items. The standards are PFA, FPO, AGMARK, ISI, etc.

As per the Prevention of Food Adulteration Act (PFA) of 1954, food adulteration is defined in the following ways:

❊   any article sold by a vendor is not of the nature, substance or quality demanded by the purchaser.

❊   any article containing any other substance which affects the nature, substance or quality.

❊   any article in which some constituent is abstracted so as to affect injuriously the nature, substance or quality.

❊ any article prepared or kept under unsanitary conditions.

❊ any article obtained from a diseased animal.

❊ any article falling below the prescribed standard.

Thus, food adulteration means:

❊ the intentional addition, substitution or abstraction of substances which adversely affect the nature of the substance.

❊ incidental contamination of quality food.

COMMON FOOD ADULTERANTS

The foods commonly adulterated in India are the following:

1. Milk

2. Milk products

3. Salted products

4. Cereals and pulses

5. Spices

6. Sweets and sweeteners

7. Edible fats and oils

8. Beverages

Table 9.1 lists the various adulterants that are present in different foodstuff.

Table 9.1   Common food adulterants

Food category	Foodstuff	Adulterants
Milk and milk products	Milk	Water, starch, defatted milk, diluted buffalo milk.
	Milk powder	Excess moisture, starch.
	Cream	Foreign fat, non-permitted colour, starch, artificial sweetener.
	Curd	Water.
	Butter, ghee	Vanaspathi, animal fat, rancid fat, etc.
Vegetable oils and fats	Oils	Argemone oil, orthotricresyl phosphate, mineral oil, hexande, rancid oil, etc.
	Vanaspathi	High-melting rancid fat, animal fat, colouring agent.
	Margarine	Excess moisture, animal fat.
Spices and condiments	Whole turmeric	Lead chromate, metanil yellow coatings.
	Turmeric powder	Lead chromate, maize powder, sand, etc.
	Garam masala, curry powder, mixed spices	Excess common salt, exhausted spices, coal-tar dyes.
	Coriander	Sulphur dioxide, woody stalks, colourants, etc.
	Cloves, cardamom	Exhausted spices and other unwanted matter.
	Chillies and powder	Oil-soluble colour, sudan I.
	Mustard	Argemone seed.
	Black pepper	Dried papaya seed.
Cereals	Wheat, rice, etc.	Stones, sand grit, insect-infected stuff.
	Atta, maida, semolina	Sand, dirt, bran, starch, talc, chalk powder.
Pulses	Dhal, black masoor	Kesari dhal, polished talc.

(Contd.)

Table 9.1   (Continued)

Food category	Foodstuff	Adulterants
Sweeteners	Sugar	White sand, iron filings.
	Bura sugar	Atta, talc, dirt.
	Ice-candy	Toxic colourant, flavours, saccharine.
	Honey	Cane sugar, invert sugar, etc.
Non-alcoholic beverages	Soft drinks	Saccharine, non-permitted colours, water.
	Coffee powder	Date-seeds, tamarind seeds, husk powder, chicory.
	Tea	Exhausted leaves, saw dust, black gram, tamarind seed powder, etc.
Alcoholic beverages	Brandy, rum, whisky, etc.	Methanol, water, isopropanol, metallic contaminants, etc.
Vegetable and fruit products		Prohibited colours, preservatives.
Miscellaneous items	Baking powder	Citric acid.
	Vinegar	Mineral acids.
	Supari	Foreign seeds or nuts.

ANALYSIS OF VARIOUS FOOD ADULTERANTS

Analysis of Adulterants in Edible Oils

Castor oil   Sample, dissolved in ether. Acidified with HCl. Few drops of ammonium molybdate are added and shaken well. Appearance of turbidity signifies the presence of castor oil as contaminant.

Sesame oil   Sample, mixed with 2% furfural and conc. HCl. Appearance of a pink colour indicates the sesame oil contamination.

Argemone oil Sample mixed with $FeCl_3$ and few drops of conc. HCl. Brown colouration signifies argemone oil as contaminant.

Linseed oil Sample, treated with bromine in CCl_4. Appearance of yellow precipitation shows the presence of linseed oil contamination.

Analysis of Adulterants in Ghee

Good quality ghee must contain small quantity of fatty acid. In a special variety the fatty acid content is up to 1.4%. General grade and standard grade varieties have 2.5 and 3% of fatty acid contents respectively. The quality is analysed by titration with acid. The Reichert–Meisel value must be less than 28.0 and Polenske value is to be 1.0–2.0 for the good quality ghee.

Analysis of Adulterants in Coffee Powder

Chicory Sample, treated with dilute acid and Saliwanoff's reagent (0.05 g resorcin in 100 mL of 4 M HCl). Red colouration indicates chicory contamination.

Tamarind powder Sprinkle the sample on a filter paper and add 5% sodium carbonate solution. Red colouration shows the presence of chicory contamination.

Analysis of Adulterants in Chilly Powder

Brick powder Sample is mixed with water and allowed to stand for sometime. The brick powder settles at the bottom.

Analysis of Adulterants in Turmeric Powder

Yellow dye The sample is mixed with alcohol and allowed to stand for some time. The alcohol layer becomes yellow in colour, immediately. This confirms the yellow dye contamination.

Lead chromate   The sample is dissolved in dilute HCl. Then, H_2S gas is passed into the solution. Black precipitation indicates the presence of lead chromate contamination.

$$Pb^2 + S^{2-} \longrightarrow PbS \text{ (black ppt.)}$$

Analysis of Adulterants in Meat

Sodium nitrite   The sample (small amount) is treated with Zn and HOAc and then with acetone. Yellow colouration confirms the presence of sodium nitrite contamination.

Note   Sodium nitrite is added as a preservative.

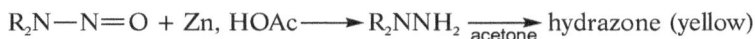

$$R_2N-N{=}O + Zn, HOAc \longrightarrow R_2NNH_2 \xrightarrow{\text{acetone}} \text{hydrazone (yellow)}$$

Analysis of Adulterants in Milk

Water   Milk sample is tested for the specific gravity. When this value is less than 1.032, then, water contamination is confirmed. The pH value of pure milk is also changed by the addition of water.

Antibiotics—Streptomycin   The milk sample is treated with dilute acid and 2,4-dinitrophenyl hydrazine. Yellow colouration indicates the presence of the adulterant, streptomycin.

HARMFUL EFFECTS OF THE ADULTERANTS

The adulterants are sure to bring about biochemical disorders, thereby physiological disorders arise. The harmful effects of various adulterants are given here.

* *Argemone oil*—Loss of eyesight, heart ailments, tumours

* *Mineral oil*—Liver damage, carcinogenic effects

* *Vanaspathi*—Liver disorders, stomach pain

* *Saw dust, colourant*—Liver disorders

* *Toxic dyes*—Carcinogenic effects

* *Resins*—Allergy, dysentery

* *Metanil yellow*—Toxic, carcinogenic

* *Washing soda*—Diarrhoea, vomiting

These harmful effects are observed in case of prolonged adulterated diet pattern. So, it is better to avoid adulterated food especially the fancy food items like coloured foods, spicy foods, synthetic beverages, etc.

Food may acquire metal contamination while preserving or cooking. The heavy metals like lead, cadmium, mercury, tin, arsenic, copper, zinc, chromium, cobalt, etc., which cause metal contamination cause harmful effects like stomach disorders, kidney malfunctioning, etc.

FOOD ADDITIVES

Food additives are the chemical substances added to food for a variety of reasons. They are sweeteners, preservatives, flavours, colourants. These additives are likely to cause health problems. These additives give no benefit at all. They are unnecessary chemicals that add commercial value to food. They have no nutritive value. So, they are considered to be adulterants as they are hazardous.

Sweeteners

They are substitutes of sugar. Examples are dulcin, saccharin, sodium cyclamate, etc. They are used in pharmaceuticals, jams, jellies, ice creams, soft drinks, etc. They cause stomach problems.

Saccharin The sample is treated with $FeCl_3$ solution. Deep red colouration signifies the saccharin contamination.

Dulcin Sample is treated with NaOH and ether. The ether layer contains dulcin. After evaporating ether, few drops of con. HNO_3 is added. Orange or brick red colouration indicates dulcin contamination.

Preservatives

They are chemical substances that preserve food from decaying by bacteria or fungi. Some are antioxidants, which prevent oxidation. Preservatives are used by companies manufacturing food products. Some common food preservatives are calcium propionate, sorbic acid, propyl gallate, sodium benzoate, nitrites, etc.

Flavours

They are chemical substances added to enhance the taste or aroma. The famous vanilla ice cream contains Vanillin. Most of the flavouring agents are esters only. For example,

* *n*-pentyl acetate gives pears flavour.
* isopentyl acetate gives banana smell.
* ethyl butyrate has strawberry flavour.
* octyl acetate bears the smell of orange.
* pentyl valerate possess the apple smell.

This way many flavouring agents are used to a variety of foodstuff, especially synthetic foods.

Colourants

These are chemical substances that are added to some foodstuff in order to give attraction. These chemicals are toxic in nature. In spite of that fact, people use these substances for certain commercial values.

Some colourants are permitted in very low dosage. Some are non-permitted as they are toxic even at very low levels. Few colourants are listed here. Lead chromate, mercuric sulphide, red lead, copper arsenite, amaranth, sunset yellow, tetrazine, black PN, food green-S, erythrosine, quinoline yellow, indigo carmine, patent blue-V.

PESTICIDE CONTAMINANTS

Liberal use of pesticides to improve the crop yield has become inevitable in the present day farming. Some of the pesticides are permanently incorporated in the human body cells/plant cells. The pesticide constituents may cause a variety of health problems. Some of the pesticides are DDT, 1,1,1-trichloro-2, 2-bis (4-chlorophenyl) ethane, malathion, parathion,carbaryl, etc.

These pesticides are extracted with $CHCl_3$ and subjected to chromatographic separation and then detected by spectrophotometric method. For example, parathion gives a broad and intense band in the UV range at 275 nm.

TOXICANTS

Toxic minerals and metallic contaminants may naturally be present in some foodstuff and water. For example, fluorine, nitrite, selenium, lead, etc. are some toxic minerals. Organic toxicants include solamine, gossypol, phalloidine, oxalic acid, anti-vitamins, cyanogenic compounds, goitrogenic compounds. These substances may cause a variety of health problems.

REVIEW QUESTIONS

Give short answers

1. Define food adulteration.
2. Give the expansion of
   i.   PFA
   ii. AGMARK
   iii. FPO
3. What are the common adulterants in milk?
4. Oils are adulterated with _____ and _____.
5. How is coriander powder adulterated?
6. Mention any two adulterants in alcoholic beverages.
7. Mineral oil is contaminated with _____ and _____.
8. Mention any two artificial sweeteners.
9. What are preservatives? Give any two examples.
10. Define the following terms:
    i. Flavouring agents
    ii. Colourants

Give detailed answers

1. Give the general principles of Codex Alimentarius.

2. Define the term "Adulteration" as per the Prevention of Food Adulteration Act (PFA), 1954.

3. List the various spices and the common adulterants.

4. List the various milk products and their common adulterants.

5. Mention the principle of analysis of the following adulterants:

    i.   Linseed oil

    ii. Argemone oil

    iii. Castor oil

6. How is chicory tested in coffee powder?

7. Mention the test to detect sodium nitrite present in meat.

8. Brief the principle of analysis of streptomycin contamination in milk.

9. Give the principle of analysis of lead chromate in turmeric powder.

10. Mention the harmful effects of the following adulterants: argemone oil, vanaspathi oil, saw dust, metanil yellow, and washing soda.

10

FOOD PRODUCTS

INTRODUCTION

This chapter discusses some of the food items and their various forms that are useful from the nutrition point of view. Although large number of foods are available and in various forms, only three kinds of food and their products are taken for discussion. They are wheat and wheat products, milk and milk products, meat, poultry and fish. Vast majority of population throughout the world use these food items. So, it would be more appropriate to consider the chemistry aspects of these food items.

WHEAT AND WHEAT PRODUCTS

Cereals are dried seeds of cultivated grasses that belong to the family Gramineae. These species include wheat, rye, barley, corn, oats, sorghum and millet. The cultivation and use of cereals have been in practice from the very early history of mankind. The civilization of Babylonia, Egypt, Greece and Rome were founded on the production of wheat, barley and millets. The cultures of India, China and Japan were dependent on rice crop.

Cereals are the cheapest sources of energy. They can be grown anywhere in the world. They give high yields and can be stored for a longer period and transported easily. Cereals are rich in protein, vitamin B, iron and phosphorus. Mostly the climatic conditions determine the type of cereals that are grown in a particular region. Table 10.1 gives the various types of cereals that are grown in different parts of the world.

Table 10.1 Cereals grown in different regions

Region	Cereals
Most parts of Asia (India to Japan)	Rice
Northern European countries	Rye
Regions of temperate climate	Wheat, corn, oats, barley, etc.
India and Africa	Sorghum
Southern Asia, Africa and Soviet Union	Millet

Classification of Wheat

Wheat belongs to the family Gramineae and the genus *Triticum*. There are more than 30,000 varieties. The important species of wheat are *T. aestivum, T. compactum* and *T. durum*. They are commercially very important. Wheat is classified into hard and soft wheat based on the endosperm. The hard variety shows the endosperm breaking in lines during milling while the soft endosperm breaks at random.

Wheat is put into five grades based on three characteristics:

    i.   minimum test weight

    ii.   maximum defects limit

    iii.   contrasting nature

Wheat Flour

Whole-wheat grains are ground to a fine powder. The grinding or milling of wheat grains is done based on the grain structure. There are a few differences in the methods adopted depending on the grain variety. The important stages of wheat flour production are as follows.

1. *Cleaning* It is done to remove the extraneous matter like straw, chaff, pebbles, bits of solids, etc.

2. *Tempering* Adding water to wheat to improve the physical state of the grain for milling.

3. *Milling* Breaking the wheat kernel by crushing between fluted rollers.

4. *Sifting or bolting* Segregating the broken particles into different size.

5. *Purification* Removal of minute bran or waste particles.

6. *Reduction* Pulverizing to desired fineness.

7. *Air classification and grading* Air is blown through the pulverized flour, which brings about separation of flour of heavier and lighter particles.

8. *Bleaching* Shaking the flour with chlorine, nitrosyl chloride or nitrogen peroxide to make it bright white.

Components of wheat flour The flour consists of the components like proteins, lipids, carbohydrates, acids and some enzymes. The chemistry aspects of these components are briefed here.

Proteins Proteins like glutenin and gliadin are present in major amounts. Albumin and globulin are present in minor amounts.

When protein fractions are mechanically agitated with water, it forms gluten. This gluten is capable of retaining gases. This property helps the flour to leaven so that the flour becomes fit for making bread.

Lipids　Lipids are about 1% and this amount varies with the variety of wheat. The lipids have little significance. The lipids are mostly glycerides, phospholipids and sterols.

Carbohydrates　Starch is about 70–75%, present in flour. It is the major constituent and forms the source of energy. During bread making, starch is acted upon by the diastase to give maltose. It contributes to the crust colour and taste of bread. Dextrins, cellulose, lignans and pentosans are minor proportions. The water-soluble simple sugars—raffinose, sucrose, maltose and dextrose—are also present in flour.

Acids　The acidity is low and varies with the type and grade of flour. If it is stored in adverse conditions, the acidity increases. The hydrogen ion concentration becomes higher for higher-grade flours and lower for lower-grade flours.

Enzymes　Amylases are present in high amounts. β-amylase is greater in proportion than α-amylase. These amylases are responsible for the conversion of starch to utilizable sugars. Lipoxidase is responsible for destroying flour pigments during the natural aging of flour.

Wheat Products

There are quite a number of wheat products available. We restrict with two kinds of products—bread and cake. The chemistry of these products is given briefly.

Bread　The wheat flour is cooked in a particular style to make a fluffy, soft and browny eatable—bread. Apart from the wheat flour, there are other ingredients added to bread in a specific formulation. The following steps are involved in bread making:

1. *Making the dough*   The flour is mixed with sufficient amount of water at 27°C to get a dough.

2. *Fermentation*   Chemical transformation of the dough at 27°C with relative humidity. About 75–80% gluten is developed.

3. *Dividing, moulding and panning*   Dough is divided into loaf-sized pieces. The pieces are moulded into loaves and placed into a pan.

4. *Pan proofing*   Panned loaves are sent to proof cabinet for the last fermentation at humidity 80–85% and at temperature 35–38°C.

5. *Baking*   The proofed loaves are sent to the oven and baked at 190–232°C. About 500 g of a loaf takes 30 minutes to get baked at 220°C.

The quality of bread is determined based on the volume, moisture, crust character, colour, texture, taste and aroma.

Cake   It is another cooked form of wheat flour. The texture is softer and more crumbly. The wheat flour should be of low protein with low α-amylase activity. The ingredients are flour, sugar, leavening agent, egg and liquid. Here, the gluten is not developed. Starch granules are made to swell uniformly. Sugar contributes to flavour, cake volume, tenderness, etc. The following steps are involved in the cake making.

1. Ingredients are mixed with beater or paddle.

2. Sugar and shortening agents are first creamed to a fluffy form.

3. Egg, leavening agent and salt are added.

4. All the ingredients are mixed uniformly to form a good blend of batter.

5. Batter is put into a pan.

6. It is then baked at a temperature of about 150°C.

Analysis of Wheat and Wheat Products

Wheat can be analysed for the following characteristics:

i. Moisture content

ii. Ash content

iii. Total carbohydrates

iv. Crude proteins

v. Crude fats

vi. Crude fibre and so on.

The methods for the aforesaid analysis are discussed in earlier chapters. The percentage composition of various constituents are given in Table 10.2.

Table 10.2   Constituents of wheat

Constituents	Amount in percentage
Moisture	13–14
Ash	15
Crude protein	8–22
Crude fat and fibre	

MILK AND MILK PRODUCTS

Milk is the normal secretion of the mammary glands of the mammals (female) for the young ones. It is defined as the lacteal secretion practically free from colostrum, obtained by the complete milking of one or more healthy cows, which contains about 8.25% milk solid-not-fat and 3.25% milk fat. The major components of milk are lactose sugar and casein protein. There are other constituents like mixture of glycerides, calcium phosphate, vitamins, enzymes, pigments and lactic acid.

Composition of Milk

Substances like water (87%), fat (3.7%), lactose (4.9%), proteins (3.5%) and ash (0.7%) constitute milk. The percentage composition varies with individual cows on such aspects as breed nature, age, lactating period, feeding, etc. The morning milk is richer in fat by 2% than evening milk. The milk obtained for the first few days after calving is known as colostrum. The composition of colostrum is entirely different from the normal milk. When the daily production of milk is decreasing, there is some increase in the percentages of fat and casein. The percentages of fat, protein and water vary with milk. But, the percentage of lactose remains fairly constant. Table 10.3 gives the percentage composition of various nutrients of milk.

Milk Grades

Milk is graded from the standpoint of bacterial count, purity, cleanliness, etc. Generally, two classes of market milk are available. One is raw milk and the other is pasteurized milk. Apart from these two classes, there are other grades of milk.

    i.   Grade A raw milk

    ii.   Grade A pasteurized milk

    iii.   Certified milk

Table 10.3   Percentage composition of nutrients in milk

Nutrients	Amount in percentage	% Chemical constituents
Water	87	
Fat	3.7	12.5% glycerol, 85.5% fatty acids
Sugars	4.9	Lactose sugars of α and β-forms
Proteins	3.5	20% whey proteins—albumins, globulins, peptones, etc.
		80% caesins of α, β and γ-forms
Minerals	0.7	22.4% calcium, 25.7% phosphorus
		26.3% potassium, 9.0% sodium
		2.6% magnesium, 0.2% iron
		14.4% chloride, 2.7% sulphite
Others	0.2	

(Source: National Research Council, USA, Washington (1953). Publication No. 253 (Composition of milks)).

Some Commercial Milk Products

The legal definitions and standards of identity for milk and milk products are not strictly uniform. However, the general descriptions and various types of milk and milk products are given in a brief way.

Raw milk   It is lacteal secretion, free from colostrum. It is obtained by complete milking of cows. It contains about 8.25% milk solids-not-fat and about 3.25% milk fat. These fat contents are always more and not less than the mentioned levels.

Pasteurized milk   Milk is heated to a temperature of about 63°C for minimum 30 minutes or 75°C for 15 seconds, then cooled to 10°C or lower. This is done using the equipment, which is properly operated and approved by health authorities.

Homogenized milk Bigger fat globules are subdivided into smaller ones by mechanical shaking. The homogenized milk should not show visible cream separation after 48 hours of storage at 7°C. The difference in fat percentage between milk (of the same lot) after boiling and that not being bottled should not be more than 10%.

Vitamin D milk Vitamin D potency is enhanced by any approved method to at least 400 USP/qt.

Soft-curd milk Milk with a curd tension of 30 g or less is called soft-curd milk. Cow milk gives a curd of greater toughness (high tension) compared to human milk. It is considered that the soft-curd milk is more digestible and preferable for infant feeding.

Skim milk This is the milk from which sufficient milk fat is removed so that it contains milk fat in the range of 0.5–2%.

Cream This is the fatty liquid separated from milk with or without the addition of skim milk, and contains not less than 18% milk fat. Cream is separated from the rest of the milk portion by bringing it to the surface on standing or by centrifuging.

Whipping cream It contains not less than 30% milk fat. The milk fat ranges between 30 and 36 per cent.

Sour cream/cultured sour cream Fluid or semi-fluid cream resulting from the souring by lactic acid-producing bacteria or a similar culture of pasteurized cream. It contains 18% milk fat and 0.5% acidity (lactic acid).

Butter It is obtained by churning the cream to granules of butter. Then, the granules are forced to a compact mass.

Buttermilk   A fluid product obtained during the manufacture of butter from curd or cream. Most of the buttermilk, used now, is the cultured buttermilk. It is obtained by inoculating the pasteurized skim milk or low-fat milk with lactic acid-producing culture, and incubating at 21°C for 12–15 hours. The curdled milk is cooled to 4°C, with agitation, to produce a smooth product. It is most stable at acidity 0.7–0.9% (lactic acid) at a storage temperature of 0.6°C.

Evaporated milk   It is the milk from which water is evaporated. The milk fat and the total solids are 7.5% and 25.5% respectively. Evaporated milk contains vitamin D (24 IU) and vitamin A (125 IU). Safe and suitable emulsifiers and stabilizers are added to this milk to prevent spoilage.

Sweetened milk   Milk is evaporated along with suitable nutritive sweeteners (sucrose + dextrose). The finished food contains 8.5% milk fat and 28% total milk solids.

Dried milk products   They are prepared from milk, skim milk, cream, whey, buttermilk and other liquid milk products. These products are dried by spray or drum drying method. The fried product is ground well to make it into a fine powder.

Ice cream   It is a frozen dairy product, made from milk products in combination with a stabilizer, sweetener and flavour. It sometimes includes colouring matter, fruit, nut, etc. The milk products in making ice cream are: cream, milk, butter, butter oil, concentrated milk, evaporated milk, sweetened milk, dried milk, skim milk, sweet cream buttermilk, condensed sweet cream buttermilk, concentrated cheese, whey and others. The stabilizers are gelatin, egg, sodium alginate, gum acacia and others. Sweetening agents are sugar, dextrose, corn syrup, maple syrup, maple sugar, honey, brown sugar, molasses, etc.

Cheese It is made by modifying or ripening the separated curd with rennet, lactic acid or other suitable enzymes or acids. There are about 60 different kinds of cheese available.

Analysis of Milk and Milk Products

The milk and milk products are analysed for the total solids, ash content, crude fat, total acidity, total protein, lactose sugar, minerals, etc. The principles of analyses of some of the constituents have been discussed in earlier chapters. Some methods are discussed here.

Determination of total solids A known amount of white sand is taken in an aluminium or porcelain dish. It is heated to 100°C. Then, a known amount of the milk sample is added to the sand in hot condition. The dish is heated on a steam bath for 10–15 minutes and kept in air oven at 110°C for 30 minutes. The dish is cooled and the weight is noted. The calculation is as follows:

Weight of Al dish + sand = W_1 g

Weight of dish + sand + milk (before heating) = W_2 g

Weight of dish + sand + milk (after heating) = W_3 g

Weight of milk = $W = W_2 - W_1$ g

Weight of solid matter = $W_s = W_2 - W_3$ g

Percentage of total solid = $(W_s/W) \times 100$

Determination of total acidity A known quantity of milk is diluted with equal quantity of water (CO_2-free). It is then titrated with a standard NaOH (N_{Na}) using phenolphthalein indicator. Then, the percentage of acidity in terms of lactic acid is calculated.

1 mL of 0.1N NaOH = 0.090 g of lactic acid

V mL of N_{Na} NaOH = $[(0.090)V \times N_{Na}]/0.1 = W_L$ g

Percentage acidity = $\{W_L / W\} \times 100$

Determination of lactose Definite amount of the sample is taken in 100-mL and 200-mL standard flasks separately. About 20 mL of mercuric nitrate solution and 15 mL phosphotungstic acid are added to each of the flasks. Each flask is made up to the required volume. Each of the solutions in the two standard flasks is taken for polarimetric measurement separately.

Reading for the first solution = S_1

Reading for the second solution = S_2

Correction factor = E (to justify any precipitation)

$2 S_2 = S_1 - 0.5E$

$E = 2 S_1 - 4 S_2$

$S_C = S_1 - E = S_1 - (2S_1 - 4S_2) = 4S_2 - S_1$

MEAT, POULTRY AND FISH

This covers the broad area of all non-vegetarian food items. Mostly these food items are protein- and fat-rich. The proteins are fibrous and globular. The lipid contents are saturated and unsaturated ones of higher order. The meat and meat products, fish and fish products are discussed briefly in this chapter.

MEAT

It is defined as the flesh of cattle, swine, sheep or goat that is fit for eating. It is made of fibres held together by connective tissues and interspersed with nerves and blood vessels. A muscle fibre comprises a number of long, thin, cylindrical rodlike myofibrils and specialized network of tubules—sarcoplasmic reticulum.

Myofibrils are bathed in an aqueous liquid called sarcoplasm. It contains 75–80% water and other cell components—mitochondria, enzyme, glycogen, adenosine triphosphate, creatine phosphate, myoglobin, etc. Each myofibril consists of two sets of filaments.

    i.   Set of thick filaments containing myosin (a protein)

    ii.   Set of thin filaments containing actin (a protein)

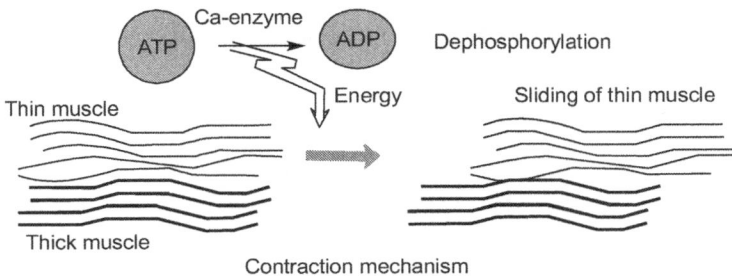

Figure 10.1   Muscle contraction mechanism

Contraction and relaxation of the muscle are attributed to the interaction between actin, myosin and ATP. Energy is released by calcium-activated enzymic dephosphorylation. The mechanism is given in Figure 10.1.

The reverse mechanism occurs when muscle relaxation takes place.

The conversion of muscle to meat involves a number of biochemical and biophysical changes occurring in three stages.

    1.   *Prerigor,* in which the muscle exists as soft and pliable (immediately after the death of animal). When the post-mortem continues the following changes occur.

* Glycogen changes to lactic acid
* pH reaches the point 5.5
* Creatine phosphate decreases
* ATP decreases
* Capacity to resynthesize ATP decreases

2. *Rigor mortis,* slow association of myofibrillar proteins (actin and myosin) to form actomyosin. This causes the inextensible properties to muscles. Muscle ceases to be elastic.

3. *Postrigor* is the resoftening of muscle from the hard state and then the slow beginning of decay.

Composition of Meat

Meat is composed of water (56–72%), protein (15–22%), fat (5–34%), soluble non-protein part (3–4%) and others. The non-protein part includes carbohydrates, inorganic salts, soluble nitrogenous substances, trace metals, vitamins, etc. The composition may vary depending on the species, breed, age, sex and the diet pattern.

Table 10.4 gives the percentage compositions of different meat varieties.

Table 10.4 Percentage composition of the constituents of meat

Type	Meat	Water	Protein	Fat	Ash	Energy (Cal)
Beaf	Chuck	56	18	25	0.8	303
Veal	Chuck	70	19	10	1.0	173
Pork	Ham	57	16	27	0.7	308
Lamb	Leg	66	18	15	1.4	209

Grades of Meat

Meat grades are the indication of meat quality. There are different levels of grading for different meat. Some have eight, six or four grades. The grades of meat are given in Table 10.5.

Table 10.5 Grades of meat quality

Grade	Beef	Veal	Pork	Lamb	Mutton
01	prime	prime	No.1	prime	prime
02	choice	choice	No.2	choice	choice
03	good	good	No.3	good	good
04	standard	standard	No.4	standard	standard
05	commercial	utility	--	cull	cull
06	utility	Cull	--	--	--
07	cutter	--	--	--	--
08	canner	--	--	--	--

POULTRY

The meat of some birds comes under the category of poultry. Most poultry are purchased in the ready-to-cook form. The feather and viscera are removed and carcass is washed while dressing the poultry. It is also a nutritious food variety.

Composition

The water content is in the range of 55–75%. Ash content is about 1%. The protein is 16–20% and the fat content ranges between 5 and 30%. The exact composition of different poultry are given in the Table 10.6.

Table 10.6   Percentage composition of poultry constituents

Type	Percentage composition				Energy
	Water	**Protein**	**Fat**	**Ash**	
Chicken (fryer)	76	19	5	0.8	124
Chicken (rooster)	63	18	18	0.9	239
Chicken (hen and cock)	57	17	25	0.9	298
Duck	54	16	29	1.0	326
Goose	51	16	32	0.9	354
Turkey	64	20	15	1.0	218

Analysis of Poultry

The poultry is analysed for the constituents like water content, protein, fat, minerals, etc. The methods of analyses are discussed in the earlier chapters. Considering the principles, the techniques are suitably adopted.

Estimation of nitrite A maximum of 200 ppm $NaNO_2$ is permitted in any meat product. It is used as a preservative. Usually, meat is tested for the presence of nitrite. The principle is briefed here.

A known amount of meat is taken in a beaker, to which sufficient amount of water is added and heated to 80°C. After cooling, the water is collected in a standard flask. The meat piece is subsequently washed with water several times. The washings are collected in a standard flask. Then, it is made up to a desired volume. 1 mL of this made-up solution is mixed with 2.5 mL NED reagent (1% N-(1-naphthyl) ethylenediamine). The total volume is made up to 10 mL. Then, the percentage transmittance (%T) of the solution is measured at 540 nm. Similar measurements are made with standard sodium nitrite solution. With the readings, a standard graph is plotted between concentration of standard nitrite solutions on the X-axis and the %T on the Y-axis. Then, from the graph, the concentration of nitrite in the sample solution is found out.

SEA FOODS

There are two major groups of fish, considered seafood even though sea has innumerable living things. The two major groups are finfish and shellfish. The finfish variety includes whole, round, drawn, dressed, etc. The shellfish variety are soft-bodied, protected by a shell, e.g. oyster, clam, scallop, etc. and some have a segmented shell, e.g. lobster, shrimp, crab, etc. There are three grades of fish—A, B and C. Grade A is of best quality and uniform size. Grade B fish are of good quality and not uniform in size. Random-sized, good quality fish are put in Grade C.

Composition

Water, protein, fat and minerals are the major constituents of fish. The proximate analysis and the energy values are given in the Table 10.7.

Table 10.7 Percentage composition of major nutrients of fish

Type	Percentage composition				Energy
	Water	**Protein**	**Fat**	**Ash**	
Black/cat fish	79	19	1.2	1.2	93
Fresh water fish	78	18	3.0	1.3	103
Cod	81	18	0.3	1.2	78
Salmon	64	23	13.4	1.4	217
Trout	78	19	2.0	1.2	101

Analysis of Fish

The fish are analysed for the constituents like water content, protein, fat, minerals, etc. The methods of analyses are discussed in the earlier chapters. Considering the principles of analysis of the constituents, the techniques are suitably adopted.

REVIEW QUESTIONS

Give short answers

1. Define cereals. Give two examples.

2. Wheat belongs to the family _____ and the genus _____.

3. Give the percentage composition of moisture, ash, protein, fat and fibre.

4. How is milk defined?

5. What are the components of milk?

6. Mention the two kinds of market milk.

7. What is skim milk?

8. How is butter obtained?

9. Brief the method of getting pasteurized milk.

10. Evaporated milk contains the vitamins _____ and _____.

11. What are the two kinds of filaments that a myofibril has?

12. Give the composition of meat.

13. Give the number of grades in each of the following meat items:

    i. Veal

    ii. Pork

14. Give three examples of finfish.

15. Brief the three grades of fish.

Give detailed answers

1. Give the various stages of production of wheat flour.

2. Briefly write about the following constituents of wheat:

    i. Carbohydrates

    ii. Enzymes

3. Brief the steps involved in bread making.

4. How is the total solid in milk determined?

5. Briefly explain the method of estimation of total acidity in milk.

6. How is lactose in milk estimated by polarimetric method?

7. Accurately 5.0211 g of milk is mixed with water to make the total volume to 10 mL. It is homogenized and titrated with 0.00398N NaOH using phenolphthalein indicator. The titration value is 9.8 mL. Find out the percentage acidity.

Ans: 0.7

8. Briefly explain the mechanism of muscle contraction.

9. Give the percentage composition of different types of poultry.

10. How is nitrite in meat estimated?

11

ENERGY METABOLISM

INTRODUCTION

Energy is the basis for all forms of life and life-related activities. Biologically, it is the "power" of an organism to do its work. In Food Science, it is the basic question—how does human body transform the elements of food into energy? Thus, the study of energy metabolism is very important in order to find a very pertinent answer to this question.

ENERGY AND METABOLISM

The term "Energy" is derived from the Greek word *energon*, which means "active" (*en* = in, *ergon* = work). Energy is the power or force that enables the body to carry on life-sustaining activities. Death is the fullstop to that activity. The word "Metabolism" has its root from the Greek word *metabole* that means "change". Metabolism is the total of all the biochemical changes that occur in the body. The substances present initially in the food are changed into other substances. The energy metabolism deals with how energy is involved in the multiple changes in the form of physiological constituents.

MEASUREMENT OF ENERGY

The body performs work based on the energy released. The energy released is in the form of heat. The heat is measured in calorie or kilocalorie (Note: This unit is in convention only in Food Science). One kilocalorie is the amount of heat required to raise the temperature of 1 kg of water by 1°C. The calorific values of various foods are measured using the bomb calorimeter. The given experiment is applied to measure the calorific value.

Experiment   A weighed amount of food is placed into the core of the calorimeter and the instrument is immersed in water. The food is then ignited by an electric spark in the presence of oxygen and burnt. The raise in the temperature of surrounding water is noted. Then, the energy is calculated. Based on this study, the following data are obtained.

  ❋   1 g carbohydrate yields 4 calories.

  ❋   1 g fat gives 9 calories.

  ❋   1 g protein is able to produce 4 calories.

The Systeme International d'Unite (SI) in 1960 adopted "Joule" or "kilojoule" as the unit of energy. It is in convention in physical sciences. Roughly, one joule is $\frac{1}{4}$ time a calorie. 1 calorie = 4.181 joules.

ENERGY CYCLE

The first law of Thermodynamics says that energy cannot be created or destroyed, but can be transformed from one form to the other. In the human body energy is available in four basic forms—chemical, electrical, mechanical and thermal energy (Figure 11.1). The energy is constantly cycled through these forms. Sun is the ultimate source of energy, which gives out energy on a very large-scale, by the non-stop nuclear fusion reactions.

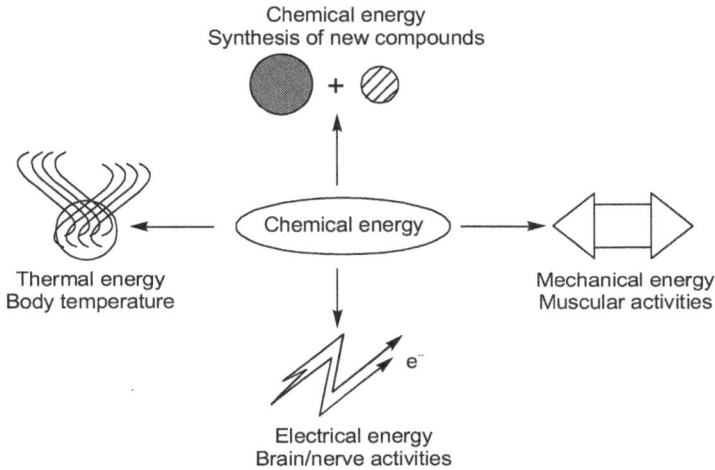

Figure 11.1 Simple scheme of conversions of chemical energy to other forms of energy

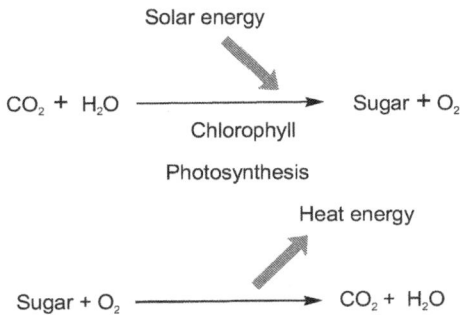

Figure 11.2 Simplified chemical reactions of synthesis and burning of sugar molecules

Plants transform the energy of the sun and store it in the form of chemical substances. The most likely chemical form is the "sugar". The synthesis of sugar molecules from carbon dioxide and water utilizing solar energy in the presence of chlorophyll, is termed photosynthesis.

The sugar molecules are supposed to be macromolecules. These bigger molecules that are present in food are converted to basic energy units—glucose. This glucose molecule is burnt with oxygen and energy is released. The reverse of photosynthesis occurs. This process is illustrated in Figure 11.2.

TRANSFORMATION OF ENERGY

Transformation of energy from its primary source (the Sun) to various other forms for biological work is shown in Figure 11.3. Metabolism is a process in which chemical form of energy is converted to other energy forms for doing work.

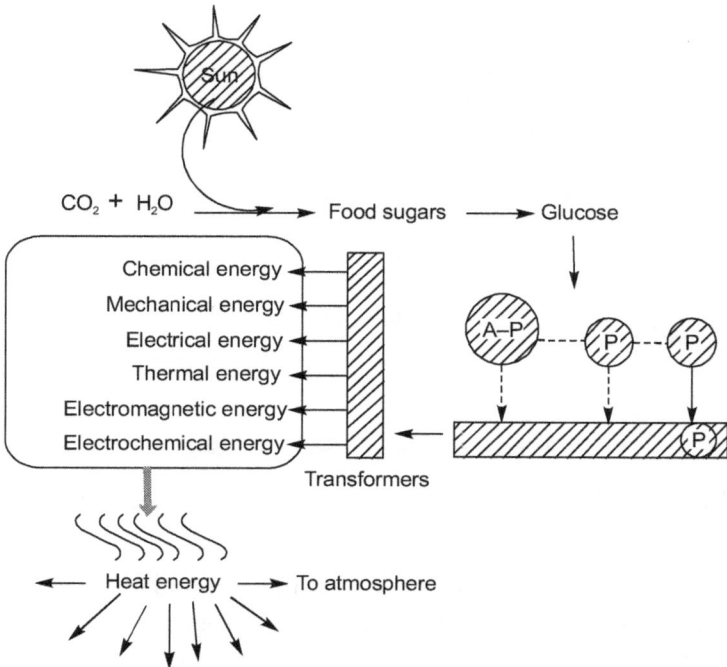

Figure 11.3   Transformation of energy from the sun

In human metabolism, the energy is always present as potential energy. It is either stored or bound in various chemical compounds (sugars, fats, and proteins). When there is an energy need for doing any work, the chemical compound is burnt and energy released is used up.

CONTROLLED ENERGY IN HUMAN METABOLISM

The energy in any system is always controlled. If it is not controlled, it may result in destruction. For example, the atom bomb while exploding releases energy that is not controlled. It results in destruction. The energy that is controlled may lead to constructive purposes. For example, atomic reactor that releases energy is controlled. It is put to use to a maximum extent.

The mechanism by which energy is controlled in the human system is mainly through chemical bonding. The chemical bonds that hold the elements together possess energy. As long as the compound is intact or stable, the energy is stored as potential energy. When the compound is broken into its parts, energy is released as the energy becomes free energy.

There are three types of chemical bonds that help storing the energy in chemical substances in the body. They are:

* Covalent bond
* Hydrogen bond
* Phosphate bond

Covalent Bond

It is formed by mutual sharing of the valence electrons present in atoms that are to be bonded. For example, water molecule is formed by the mutual sharing of valence electrons of the hydrogen, oxygen and hydrogen atoms (Figure 11.4).

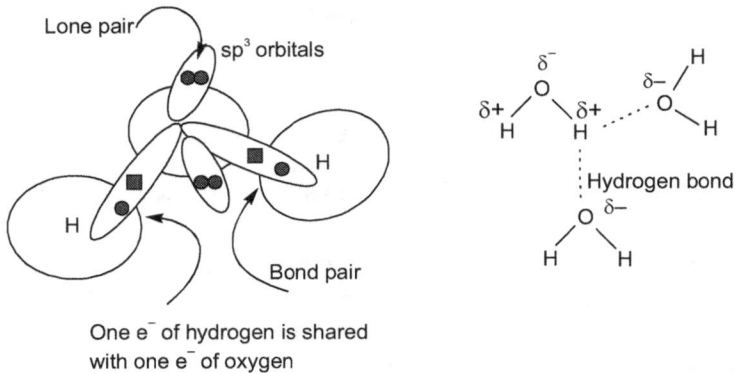

One e⁻ of hydrogen is shared
with one e⁻ of oxygen

Figure 11.4   Covalent and hydrogen bonding in water molecules

Hydrogen Bond

It is a weak bond formed due to coulombic attraction between the positively polarized hydrogen-end of one molecule and the negatively polarized oxygen-end of another molecule. It is found only in selected compounds having covalent bonds that are formed between hydrogen and the electronegative atom like oxygen, nitrogen or fluorine. The hydrogen bonding is shown in Figure 11.4. The coiled arrangement of protein molecules, nucleic acids and other macromolecules are intact and stable because of hydrogen bonding.

Phosphate Bond

The phosphate (PO_4) radical is highly labile. More energy is required to bind it than to bind carbon or most other radicals. More free energy is released when phosphate bond is broken.

Phosphate bonds are referred to as high-energy bonds. For example, adenosine triphosphate (ATP) breaks down to give adenosine diphosphate and a phosphate group. In this reaction, energy is released.

$$A\text{-}PO_4\sim PO_4\sim PO_4 \quad \rightarrow \quad A\text{-}PO_4\sim PO_4 + PO_4$$

Sometimes low-energy phosphate bonds are formed. For example, the phosphorylation of glucose.

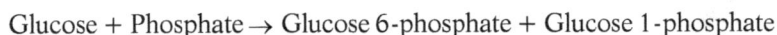

Glucose + Phosphate → Glucose 6-phosphate + Glucose 1-phosphate

This activated glucose is involved in the cell metabolism. Later, in the total process of oxidation, glucose phosphate high-energy bonds are formed.

CONTROLLED REACTION RATES

In the metabolic processes, several chemical reactions that involve energy take place. These reactions are to be controlled. If these reactions are not controlled, it may take several years for their completion or the reactions may be uncontrollably fast. The enzymes, coenzymes and hormones are available in the body such that the reactions occur at optimum speed.

Enzymes

They are biochemical catalysts. There are thousands and thousands of enzymes. But, each enzyme is very selective and specific in its action. The nature of the enzyme-catalysed biochemical reactions are "lock-and-key mechanism". (Figure 11.5).

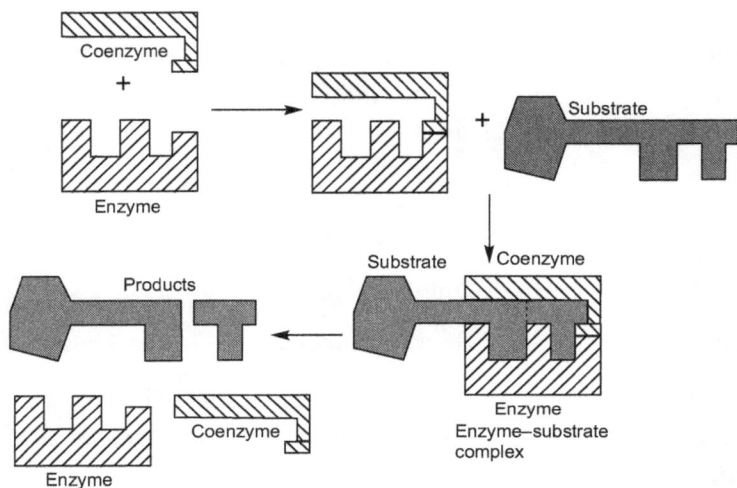

Figure 11.5 The lock-and-key mechanism of enzyme-catalysed reactions

Coenzymes

They are activators of enzymes. Each enzyme has its coenzyme either attached or detached, before the enzyme involves in a biochemical reaction. The coenzyme may be an organic group, a metal or an inorganic part. For example, enzyme dehydrogenase has it coenzyme niacin. The flavoprotein system has riboflavin as the coenzyme.

Hormones

The word hormone is derived from the Greek word *hormaein*. It means "set in motion". Hormones are secretions of endocrine glands and they perform regulatory functions. For example, the rate of oxidative reactions in tissues is controlled by thyroxine from thyroid gland which in turn is controlled by thyrotropic hormone from the anterior pituitary gland.

TYPES OF METABOLIC REACTIONS

There are two types of metabolic reactions going on in energy metabolism. They are

    i.   anabolism

    ii. catabolism.

Each process requires energy, thus the free energy decreases in each process.

Anabolism

It is a process in which simple molecules are used to synthesize complex substance. More the complexity of the substance that is synthesized, more is the energy bound to it.

Simple molecules → Complex molecules

Catabolism

It is a process by which bigger molecules are broken down to give simple substances and free energy is given out.

Bigger molecules → Simple substances + Free energy

Free energy is constantly released that leads to the deficit of energy. This deficit is compensated by the supply of food. When food is not supplied (during starvation/fasting), the body draws energy from its own store.

   ※   12–48 hours reserve glycogen is available in liver and muscle.

   ※   Energy stored as protein in limited amounts in muscle mass, but in greater volume than glycogen stores.

✳   Storage in adipose tissue is unlimited. This stored fat provides needed energy. But, it varies from person to person.

ENERGY METABOLISM

Basal Metabolism

It is the measure of energy produced in the maintenance of the body at rest after a 12-hour fast. The basal metabolic rate (BMR) is the rate of internal chemical activity of resting tissue. The various body tissues contributing to the BMR is given here. Brain, liver, gastrointestinal tract, heart and kidney together contribute to 5% of the total body weight. It is 60% of the total basal metabolic processes.

Methods of Measuring BMR

There are two methods of measuring the BMR using calorimeter. One is the direct calorimetric method and the other is the indirect calorimetric method.

In the direct calorimetric method, the person is allowed to get into a large chamber. The heat produced in his body is measured. This instrument is costly and so it has limited use. It is used only in research centres.

The indirect calorimetric method measures the exchange of gases in respiration when a person is at rest. The calories are computed as the average of the amount of oxygen consumed during two 6-minute periods.

RQ is the equivalent to heat given off by the body. BMR is the amount of calories given off per hour per square metre of the body surface area, with corrections of age, sex, height and weight. There are conditions necessary for the accurate measurement of BMR.

* Person must be on fasting 12 hours before the study.

* Person should be in a relaxed state with at least 2 hours bed rest.

* Person should be recumbent during test.

* Person is expected to be fully awake.

* Room temperature to be 20–25°C.

$$\text{Respiratory quotient (RQ)} = \frac{\text{Volume of } CO_2 \text{ produced}}{\text{Volume of } O_2 \text{ consumed}}$$

Factors Influencing BMR

* During the growth period, BMR is higher.

* Smaller persons have higher BMR that the larger person.

* Women have lower BMR than men.

* BMR increases during pregnancy (20–25% increase).

* BMR rises during lactation period.

* Fever increases the BMR by 1% for each 1°F rise.

* BMR increases with decrease in atmospheric temperature.

* Diseases involving higher cellular activity increase BMR.

* Starvation and malnutrition lower the BMR.

* Thyroxine stimulates BMR.

Other Influences on Calorie Requirement

❋   Muscular work makes the calorie requirement more.

❋   Mental efforts demand more calories of energy.

❋   Emotional state requires more energy.

❋   Food intake increases the expenditure of calories for digestion and absorption.

TOTAL ENERGY REQUIREMENT

The number of calories necessary to replace daily basal metabolic loss plus loss from exercise and other activities is called total energy requirement. It is given in the Tables 11.1 and 11.2.

Table 11.1   Energy spent per hour during different activities for a man (70 kg weight)

Activity	Cal / h	Activity	Cal / h
Sleeping	65	Slow walking	200
Awake, lying still	77	Active exercise	290
Sitting at rest	100	Severe exercise	450
Relaxed standing	105	Swimming	500
Dressing	118	Running	570
Sewing	135	Fast walking	650
Typewriting	140	Walking upstairs	1110
Light exercise	170		

Table 11.2   Daily energy requirement

Group	Age (Yr)	Weight (kg)	Height (cm)	Energy (kcal)	Energy (kJ)
Infants	0–0.5	6	60	kg × 117*	kg × 489.5**
	0.5–1.0	9	71	kg × 108*	kg × 451.9**
Children	1–3	13	86	1300	5439.2
	4–6	20	110	1800	7531.2
	7–10	30	135	2400	10041.6
Males	11–14	44	158	2800	11715.2
	19–22	67	172	3000	12552.0
	23–50	70	172	2700	11296.8
	51+	70	172	2400	10041.6
Females	11–14	44	155	2400	10041.6
	19–22	58	162	2100	8786.4
	23–50	58	162	2000	8368.0
	51+	58	162	1800	7531.2
Pregnant				+ 300	+ 1255.2
Lactating				+ 500	+ 2092.0

*   Kilogram weight of the infant multiplied by 117 kcal or 108 kcal.
**   Kilogram weight of the infant multiplied by 489.5 or 451.9 kj.

REVIEW QUESTIONS

Give short answers

1. Define calorie. The energy 1 kcal is equal to _____ kJ.

2. Give the amount of energy in kJ that is produced by each of the following substances:

    i.   Carbohydrate

    ii. Fat

    iii. Protein

3. Write briefly the chemistry of photosynthesis.

4. What are the four different forms of chemical energy in human metabolism.

5. What are energy controls in metabolism.

6. Brief the covalent bonding in water.

7. Explain the hydrogen bonding in water.

8. What are enzymes? Give its types.

9. Mention the coenzymes in the following enzymes:

    i.   dehydrogenase

    ii. flavoprotein

10. What do you mean by the term anabolism?

11. Briefly explain the basal metabolism.

12. What is respiratory quotient (RQ)?

13. Give the energy spent per hour for a man during the following activities:

    i.   sleeping

    ii.   fast walking

14. Give the energy requirement in kJ for the following:

    i.   Children [7–10 years], Females [23–50 years].

    ii.   What are influences on the caloric requirement?

Give detailed answers

1. Brief the principle of measuring calorific value of a food by bomb calorimetry.

2. Give the scheme of transformation of solar energy to various forms of living things.

3. Write about the energy controls in metabolism.

4. Briefly explain the lock- and-key mechanism of enzyme reactions.

5. Write about the controls of reaction rates in metabolism.

6. Write about the types of metabolic reactions.

7. Expand the term BMR. How is it measured?

8. Mention the factors influencing the BMR.

APPENDIX

EASY-TO-DO HOME KIT FOR FOOD-ADULTERATION TESTING

This home kit is very easy to handle and the tests can be carried out without much difficulty. These tests are sure to show clear results and the simple and commom food adulterants can be tested.

Reagents Required

Solution 1 (S1)   10 g of ferric chloride is dissolved in 100 mL of water and homogenized to get a clear solution.

Solution 2 (S2)   Ether or pet ether (about 100 mL) is taken in a bottle.

Solution 3 (S3)   Concentrated hydrochloric acid (about 100 mL) is taken in a bottle.

Solution 4 (S4)   10 g of potassium hydroxide is dissolved in 100 mL of ethyl alcohol and homogenized to get a clear solution.

Solution 5 (S5)   5 g of sodium carbonate is dissolved in 100 mL of water and homogenized to get a clear solution.

Solution 6 (S6)   1 g of iodine crystal and 0.5 g of potassium iodide are ground well, dissolved in 100 mL of water and homogenized to get a clear solution.

Solution 7 (S7)   Ethyl alcohol (100 mL) is taken in a bottle.

Solution 8 (S8)   2 g of antimony chloride is dissolved in 100 mL of chloroform and homogenized to get a clear solution.

Apparatus Required

Spirit lamp

Wash bottle

Spatula

Glass rod

Funnel

Watch glass

Test tube stand

Test tube holder

Test tubes

Boiling tube

Tripod stand

Cleaning brush

Filter paper

S.No.	Food item	Adulterant	Test	Observation	Harmful effects
01	Mustard oil	Argemone oil	0.5 mL of mustard oil + 1 mL of **S1** + 1 ml of dilute **S3**.	Formation of needle-shaped brown crystals confirms the presence of argemone oil.	Vision defects Heart disease Tumour
02	Edible oil groundnut/coconut oils	Mineral oil	0.5 mL of sample + 0.5 mL of **S4**. Warm in a water bath for some time. Add 1 mL of water.	Turbidity shows the presence of mineral oil.	Liver damage Carcinogenic effects
		Karanjia oil	0.5 mL of sample + 0.5 mL of **S8**. Shake well.	Appearance of orange or yellow colour, immediately.	Heart problems
		Castor oil	0.5 mL of sample + 0.5 mL of **S2**. Warm in a water bath for some time. Cool in ice.	White turbidity confirms presence of castor oil.	Stomach problems
03	Ghee/butter	Vanaspathi	0.5 g of sample + 0.5 mL of **S3** + 0.5 g of sugar. Shake vigorously. Allow it to stand for 5 minutes.	Appearance of crimson red confirms the presence of vanaspathi.	Liver disorder Stomach pain
		Smashed potatoes and other starch substances	0.5 g of sample + 1 mL of **S6**.	Blue colouration indicates the presence of starch.	Stomach disorder

S.No.	Food item	Adulterant	Test	Observation	Harmful effects
04	Coffee powder	Chicory powder	0.5 g of sample + 1 mL of water.	Red colour separation indicates the presence of chicory.	Stomach disorder
		Tamarind powder	Sprinkle 1 g of sample on a filter paper + 1 mL of **S5**.	Red colouration indicates the presence of tamarind powder.	Stomach disorder
05	Tea dust	Used tea dust Saw dust Some colourants	Sprinkle 1 g of sample on a wet filter paper.	Colour separation indicates the presence of non-permitted substance.	Liver disorder
06	Dhal or lentils	Kesari dhal Toxic dyes	View at the lentils using magnifying lens.	Shape of the kesari dhal is different from lentils.	
			0.5 g of sample +1 mL of **S3**. Keep in water bath for few minutes.	Pale red colouration indicates the presence of kesari dhal.	Toxic dyes are carcinogenic.
07	Chilli powder	Brick powder	0.5 g of sample + 1 mL of water.	Brick-red sand settles down.	
			0.5 g of sample + few drops of **S3**. Make a paste and keep it in flame.	Brick-red flame indicates the presence of calcium.	

S.No.	Food item	Adulterant	Test	Observation	Harmful effects
08	Turmeric powder	Yellow aniline dye	0.5 g of sample + 1 mL of water +1 mL of **S7**. Shake well. Allow it to stand for sometime.	The solution turns yellow immediately.	Anaemia
		Metanil yellow	0.5 g of sample + 1 mL of water + 1 mL of **S3**. Shake well. Allow it to stand for sometime.	Appearance of magenta red confirms the presence of metanil yellow.	Liver disorder Cancer
09	Asafoetida	Foreign resins Galbanum Colophony resin	0.5 g of sample + 1 mL of **S7**. Shake well. Filter. To the filtrate, add 1 mL of **S1**.	Olive-green colouration indicates the presence of foreign resins.	Allergy Stomach disorder
			0.5 g of sample. Burn it in a flame.	Burning with sooty flame signifies pure asafoetida.	
10	Milk *Khoa* *Bura*	Starch	0.5 g of sample + 1 mL of water. Boil. Cool. Add 1 mL of **S6**.	Blue colouration indicates the presence of starch.	Stomach disorder Liver damage
		Water	Allow 2 mL of milk sample to flow over a vertical surface.	Free flow without leaving a trail.	
11	Pulses (Green peas)	Dyestuff	5 g of sample + 10 mL of water. Allow it to stand for two minutes.	Green colouration of water indicates the dyestuff.	Stomach disorder Ulcer

S.No.	Food item	Adulterant	Test	Observation	Harmful effects
12	Honey	Sugar water	Dip a cotton wick in 5 mL of the sample. Then burn the wick.	If the wick does not burn or burns with cracking sound, then sugar water is present.	
13	Sugar	Washing soda	0.5 g of sample + 1 mL of water. Dip a red litmus paper.	Blue colouration indicates the presence of washing soda.	Diarrhoea Vomitting
			0.5 g of sample + 1 mL of water + 0.5 mL of **S3**.	Brisk effervescence indicates the presence of washing soda.	
14	Sweets Jams Juices	Non-permitted colourants	0.5 g of sample + 1 mL of water. Shake well. Add 1 mL of **S3**.	Pinkish-red colouration indicates the presence of metanil yellow.	Toxic Cancer-causing
15	Wheat	Ergot (a fungus containing a poisonous substance)	0.5 g of sample + 3 mL of common salt solution.	The impurity of ergot floats on top.	Poisonous
16	Black pepper	Papaya seed	Small amount of the sample is viewed through a lens.	Oval-shaped, brownish papaya seeds are smaller in size than pepper.	

S.No.	Food item	Adulterant	Test	Observation	Harmful effects
17	Cloves	Any stick having the shape of a clove dipped in volatile oil extracted from cloves	Small amount of the sample is viewed through a lens.	Unshaped sticks are smaller in size than cloves with pungent smell.	
18	Mustard seeds	Argemone seeds	Small amount of the sample is viewed through a lens.	Mustard seeds have smooth surface whereas the argemone seeds have rough surface and are bigger in size.	
19	Jaggery	Washing soda	5 g of sample + 1 mL of water + few drops of S3. Allow it to stand for few minutes.	Brisk effervescence indicates the presence of washing soda.	Diarrhoea Vomitting
20	Common salt	Chalk powder	5 g of sample + 1 mL of water. Allow it to stand for few minutes.	The solution turns white indicating the presence of chalk powder.	

GLOSSARY

Absorption A complex process by which simple and important chemical substances of food are taken up by the cells. The nutrients are dissolved in the lipid of the cell membrane and then diffused into the cells.

Acidulants Additives that give a sharp taste to foods. They also assist in the setting of gels and act as preservatives.

Amino acids A type of organic compounds having two functionality groups—COOH and NH_2 groups. There are twenty-one amino acids known.

Amphoteric character Proteins being able to react with both acids and bases as they possess both acidic and basic nature.

Anabolism A constructive process in which new substances are produced from simple compounds that are obtained from food.

Anomerism The alpha- and beta-labelling of glucose and fructose arising due to change in the spatial orientation of –H and –OH at the C_1 carbon of the ring.

Antioxidants Chemical substances that prevent the oxidation of food when exposed to oxygen in the air. Oxidation of food is a destructive process, causing loss of nutritional value and changes in chemical composition. Antioxidants are added to food to slow the rate of oxidation and, if used properly, they can extend the shelf life of the food in which they have been used.

Atherosclerosis The collection of excess cholesterol in the arteries. Human body often produces more cholesterol than is needed and the excess is excreted or collected in the body as gallstones or deposited in the arteries.

Bactericidal method A method of food preservation in which most of the microorganisms are killed. Examples of bactericidal methods of preservation are cooking, canning, pasteurization, sterilization, irradiation, etc.

Bacteriostatic method A method of food preservation in which multiplication of microorganisms is prevented. This may be achieved by removal of water, use of acids, oil, or spices and by keeping at low temperature. The common methods using this principle are drying, freezing, pickling, salting, and smoking.

Baking A method of cooking using hot air. Basically, it is a dry heat method. But, steam that comes from the foodstuff is also used. For example, bread, cakes and puddings are cooked by this method.

Balance In living things, there exists a balance and there is a constant inflow and outflow of materials in building up and breaking down of parts. A balanced depositing and mobilizing of constituents is termed as equilibrium or homeostasis. The activities connected to the maintenance of this equilibrium are called homeostatic mechanisms. The balance between the body parts and functions is life-sustaining.

Basal metabolism The measure of energy produced in the maintenance of the body at rest after a 12-hour fast. The basal metabolic rate (BMR) is the rate of internal chemical activity of resting tissue.

Blanching Immersing the foodstuff first in boiling liquid and then in cold water. Foodstuff like potatoes, carrots, etc. are blanched.

Bread The fluffy, soft and browny eatable obtained by cooking wheat flour in a particular style. Apart from the wheat flour, there are other ingredients added to bread in a specific formulation.

Bulk sweeteners A group of sweeteners predominantly composed of the polyols. They are derivatives of normal sugars and exhibit carbohydrate-like structure and functionality. The polyols can often be used as direct replacements for sugar. Polyols are suitable for diabetics by virtue of their reduced glycaemic index. They are also reported to play a role in actively reducing the risk of tooth decay. These are 60% as sweet as table sugar.

Butter The product obtained as granules from churning of cream. Then, the granules are forced to a compact mass.

Cakes A cooked form of wheat flour. The texture is softer and more crumbly than bread. The wheat flour should have low protein content and low α-amylase activity. The ingredients are flour, sugar, leavening agent, egg and liquid.

Calcium An important macronutrient useful in bone formation. Deficiency causes bone weakness.

Canning A method of food preservation in which the food is boiled in the can to kill all the bacteria and sealed to prevent any new bacteria from getting in. The food in the can is completely sterile, and so it does not get spoiled.

Carbohydrates One of the nutrients present in food. They have the general formula $C_n(H_2O)_n$ where n can be in the range of 3 to 1000. Carbohydrates may be defined as poly-hydroxyaldehydes or ketones and/or their derivatives.

Carbonating A process in which carbon dioxide gas is made to dissolve under pressure. By eliminating oxygen, carbonated water inhibits bacterial growth. Carbonated beverages (soft drinks) therefore contain a natural preservative.

Catabolism A destructive process in which the complex substances are broken down into simpler compounds.

Cell The most fundamental unit of any living organism. It is just like the brick that forms the fundamental unit of a building. The survival of a cell depends on its favourable environment.

Cheese A milk product made by modifying or ripening the separated curd with rennet, lactic acid or other suitable enzymes or acids. There are about 60 different kinds of cheeses available.

Chemical preservatives The commonly used and also the most abused method of food preservation using chemical substances. Chemical preservatives serve as either anti-microbial or antioxidant or both.

Coenzymes   The activators of enzymes. Each enzyme has its coenzyme either attached or detached, before the enzyme involves in a biochemical reaction. The coenzyme may be an organic group, a metal or an inorganic part. For example, the enzyme dehydrogenase has niacin as its coenzyme.

Compound lipids   Esters of alcohol and fatty acids with one more organic group. Phospholipids, glycolipids, etc. belong to this type of lipids.

Compound proteins   Simple proteins along with some non-protein parts.

Compound sugars   Sugars other than monosaccharides including the di-, tri-, oligo-, and polysaccharides. They are hydrolysable ultimately to give monosaccharides.

Controlled energy in human metabolism   The energy in any system is always controlled for constructive purposes. If it is not controlled, it may result in destruction. For example, the atom bomb while exploding releases energy that is not controlled. The mechanism by which energy is controlled in human system is mainly through chemical bonding. The chemical bonds that hold the elements together possess energy. As long as the compound is intact or stable, the energy is stored as potential energy. When the compound is broken into its parts, energy is released as the energy becomes free energy.

Controlled reaction rates   In metabolism, there are many chemical reactions that take place involving energy. These reactions are to be controlled. The enzymes, coenzymes and hormones available in the body keep these reactions under control, such that the reactions occur at the optimum speed.

Cooking   A process in which foodstuff is changed desirably in order to make it fit for eating. Most foodstuff need to be cooked excepting few friuts, vegetables and nuts. The changes brought about by working may be physical or chemical. There are quite a number of processes employed in cooking.

Covalent bond   The chemical bond formed by mutual sharing of the valence electrons present in atoms that are to be bonded. For example, a water molecule is formed by the mutual sharing of valence electrons of the oxygen and hydrogen atoms.

Cream Fatty liquid separated from milk with or without the addition of skim milk, which contains not less than 18% milk fat. Cream is separated from the rest of the milk portion by bringing it to the surface on standing or by centrifuging.

Denaturation The drastic changes that proteins undergo in their behaviour when they are subjected to heating or reaction with strong acids, bases or other reagents. Thus, they lose their physiological functions. For example, egg turns into a solid on heating.

Derived lipids Substances obtained from simple and compound lipids. Steroids are derived lipids that are widely distributed in animals and plants. They are biologically important. Some of them are sterols, bile acids, sex hormones, cardiac aglycones, etc.

Derived proteins Polypeptides in different fragments depending upon the size. The fragments vary from small to large in size.

Digestion The process by which complex food materials are broken down into simple chemical substances that are suitable in size and composition for absorption. It consists of a series of physical and chemical changes effected on the food.

Dipolar nature of amino acids Amino acids exist in the ionic or dipolar form. It is otherwise called zwitter ionic form. The existence of this ionic form is supported by the crystalline, high-melting nature and their solubility in highly polar solvents.

Emulsifiers Chemical substances that help the mixing of oil with water in foodstuff. The emulsifier keeps these mixed, and without it the oil and water separate.

Energy The term "Energy" is derived from the Greek word *energon*, which means "active" (*en* = in, *ergon* = work). Energy is the power or force that enables the body to carry on life-sustaining activities.

Energy cycle The first law of thermodynamics states that energy can neither be created nor destroyed, but can be transformed from one form to the other. In the human body energy is available in four basic forms—chemical, electrical, mechanical and thermal energy. The energy is constantly cycled through these forms. Sun is the ultimate source of energy, which gives out energy on a very large-scale.

Energy metabolism The involvement of energy in the multiple changes in the form of physiological constituents.

Energy unit The calorific values of various foods measured using the bomb calorimeter. One kilocalorie is the amount of heat required to raise the temperature of 1 kg of water by 1°C.

Enzymes Biochemical catalysts. There are thousands and thousands of enzymes. But, each enzyme is very selective and specific in its action. The enzyme-catalysed biochemical reactions are of the nature of the "lock-and-key mechanism".

Essential amino acids Amino acids that are supplied only through food and are not synthesized by the body. Of the twenty-one amino acids, nine are essential amino acids.

Essential fatty acids The essential fatty acids are not synthesized by the body. They are obtained through diet. Absence of such essential fatty acids may cause a specific deficiency disease. For example, a type of eczema in infants is due to the deficiency of the fatty acid, linoleic acid.

Fatty acids Long chain carboxylic acids. The carbon chain may be C_{10}–C_{24}. There may be unsaturation points, one or more in number in between the long aliphatic chains. They are called unsaturated fatty acids. The fatty acids with no unsaturation are called saturated fatty acids.

Fermentation The process by which the complex substance in the food is transformed into simpler ones with the help of beneficial microorganisms. The enzymes of the microorganisms bring about this transformation. For example, leavening of the flour enhances the eatability of the food.

Flavouring agents The chemical substances that impart smell to the foodstuff. The taste of a food is a complex mixture of flavours and aroma or smell. The receptors for the human sense of flavour are located on the mucous membrane. When the molecule of a particular size fits within the cavity of the receptor, the smell is felt.

Fluorine It is a trace nutrient useful for bone formation. Deficiency leads to fluorosis and dental problems.

Food The substance that supplies the body with energy-producing materials, and that maintains the body and regulates the body processes. Thus, food plays a vital role in physiological functions.

Food additives The chemical substances added to food for a variety of reasons. They are sweeteners, preservatives, flavours, colourants. These additives are likely to cause health problems. They are unnecessary chemicals that add commercial values to food. They have no nutritive value.

Food adulterants The chemical substances intentionally added to food to increase the quantity of food thereby reducing the quality. Adding water to milk is an example.

Food adulteration The intentional addition of some unwanted substances to food or removal of valuable substances from food. This adulteration has become a menace and it poses lot of problems to the mankind.

Food colourants Characteristic chemical substances that impart colour to a food found in nature. Normally, the colour-imparting substances have many conjugated double bonds. Food colourants can also be synthetically prepared.

Food preservation The process by which food is protected from spoilage. Preservation is not a permanent protection for a foodstuff from spoilage. It is only for a short period. Preserved food is in no way superior to fresh food. But the nutrients are not lost from food due to preservation.

Food spoilage The process by which food gets deformed or decayed, and becomes unfit for eating. It is a natural process by which food undergoes physical and chemical changes to form some harmful products. These changes are caused by microorganisms (like bacteria, fungi, etc.) and macroorganisms (like rodents, insects, etc.) bring about these changes.

Freeze-drying A special form of drying that removes all moisture and tends to have lesser effect on a food's taste than normal dehydration. In freeze-drying, food is frozen and placed in a strong vacuum. The water in the food then sublimes, that is, it turns straight from ice into vapour.

Health   A state of complete physical, mental and social well-being, and is not merely the absence of disease or physical deformity.

Hormones   Secretions of endocrine glands that perform regulatory functions. The word hormone is derived from the Greek word *hormaein* meaning "set in motion". For example, the rate of oxidative reactions in tissues is controlled by thyroxine from the thyroid gland.

Hydrogenation of oils   The process of transformation of vegetable oils, which are usually liquids, to solids by removing their unsaturation positions and making them saturated. This is done for certain commercial reasons. Hydrogenation increases the shelf life, packagability, storage and transportability. But, hydrogenation makes the oil less digestible.

Hydrogen bond   A weak bond formed due to coulombic attraction between the positively polarized hydrogen end of one molecule and the negatively polarized oxygen end of another molecule. It is found only in selected compounds having covalent bonds that are formed between hydrogen and the electonegative atoms like oxygen, nitrogen or fluorine.

Hypercholesterolaemia   The condition where blood contains higher level of cholesterol ($> 250 \, mg/100 \, mL$). The normal cholesterol level in blood is $180 \, mg/100 \, mL$). Hyper-cholesterolaemia leads to coronary heart disease.   This trend is influenced by the factors like calorie intake, cholesterol intake, fat intake, and low essential fatty acid content in fat.

Ice cream   A frozen dairy product made from milk products in combination with a stabilizer, sweetener and flavourant. It sometimes includes colouring matter, fruit, nut, etc.

Intense sweeteners   Sweeteners that have a diverse range of chemical structures and are very much sweeter than sugar. Intense sweeteners are generally used in products to reduce calories or for making tooth-friendly sweets.

Intermediary metabolism The physical and chemical changes that take place in the internal environment. A number of nutrients are available in the cells at any moment for the metabolic purposes. For example, glucose, fatty acids, glycerol and amino acids can enter the common pathway that yields energy. Sometimes glucose can be metabolized to fatty acids and cholesterol, amino acids can act as the sources of glucose and fatty acids and so on.

International Codex Alimentarius Commission The principal organ of a worldwide food standard programme. It is a joint venture of the FAO and WHO.

Iodine A mineral nutrient useful for the formation of thyroxine. Deficiency of this mineral leads to goitre.

Iodine value The number of milligrams of iodine reacted with one gram of fat, signifying the degree of unsaturation in lipid.

Iron A very important nutrient for the formation of haemoglobin and ATP.

Isoelectric point The pH at which amino acids and proteins exhibit electrical neutrality.

Ketosis A condition where ketone bodies, which are intermediate products of fat metabolism, get accumulated when the conversion of ketones to fatty acids is incomplete.

Lipids A group of organic compounds—fats, oils, waxes and other related compounds—that are greasy to touch and insoluble in water. They have the constituent elements C, H, O, N, P, etc. Lipids are chemically defined as the esters of fatty acids and alcohols. The fatty acids are the long-chain aliphatic part with carboxylic acid functionality. Lipids are a concentrated source of energy for living cells and they supply two-fifths of the total calorie intake.

Magnesium One of the major nutrients important for the enzymes involved in oxidative phosphorylation.

Major minerals Nutrients that are required in large amounts. Calcium, magnesium, sodium, potassium, phosphorus, sulphur and chlorine are some examples.

Malnutrition The impairment of health due to deficiency or excess of nutrients in food. There are two types of malnutrition. One is undernutrition and the other is overnutrition.

Marinating Soaking the food in a marinade to make it tender or to have fine smell. The marinade may be a liquid prepared for a type of food. For example, the meat is marinated with a liquid made up of oil, spices and acid.

Meat The flesh of cattle, swine, sheep or goat that is fit for eating. It is made of fibres held together by connective tissues and interspersed with nerves and blood vessels. A muscle fibre comprises a number of long, thin, cylindrical rod-like myofibrils and specialized network of tubules, the sarcoplasmic reticulum.

Metabolism The total of all the biochemical changes that occur in the body. Metabolism involves release of nutrients from food, transformations of the nutrients required for the energy release, synthesis of regulatory materials, etc. It consists of two processes one is anabolic (constructive) process and the other is a catabolic (destructive) process.

Milk The normal secretion of the mammary glands of mammals (female) for the young ones. It is free from colostrum, and it contains about 8.25% milk solid-not-fat and 3.25% milk fat. The major components of milk are lactose sugar and casein protein.

Mutarotation The change in the optical rotation of glucose when it is kept in solution form, due to the interconversion between the α and β-forms.

Negative nitrogen balance A decay process in which the removal of protein is more than the intake.

Niacin/Nicotinic acid A vitamin discovered by Joseph Goldberger in 1900, which is a definite cure for pellagra and other related diseases. In 1911, Casimir Funk (London) isolated nicotinic acid from rice polishings.

Non-essential amino acids Amino acids that can be synthesized by the body. There are twelve non-essential amino acids.

Non-reducing sugars Sugars that do not react with Fehling's and Tollen's reagents.

Nutrients The basic constituents of food that must be supplied to the body in appropriate amounts. There are six different nutrients—carbohydrates, proteins, lipids, vitamins, minerals and water.

Nutritional status The condition of health of an individual influenced by the utilization of nutrients. It is determined by correlating the information on the medical and dietary history of individuals with medical examination reports.

Over-nutrition The excess of calories and/or the excess of one or more nutrients.

Peeling and stringing Removal of the outer skin of a fruit, vegetable or nut. Thus, the non-edible portion or toxic substances are removed. In some cases, the useful nutrients are also removed in this process.

Pesticide contamination Liberal use of pesticides to improve the crop yield has become inevitable in the present-day farming and this has led to the permanent incorporation of various constituents of the pesticides in the human body

cells/plant cells. This is called pesticide contamination which may cause a variety of health problems.

Phosphate bond High-energy bonds, which when broken, release a great amount of free energy. The phosphate (PO_4) radical is highly labile. More energy is required to bind it than to bind carbon or most other radicals. For example, adenosine triphosphate (ATP) breaks down to give adenosine diphosphate and a phosphate group. In this reaction, energy is released.

Phosphorus An important macronutrient useful in the formation of bone and nucleic constituents.

Pickling A widely used method to preserve meat, fruits and vegetables in the past, but today it is used almost exclusively to produce pickles, or pickled cucumbers. Pickling uses the preservative qualities of salt, combined with the preservative qualities of acid, such as acetic acid (vinegar).

Poaching Cooking in minimal quantity of water at a temperature of 80–85°C. The temperature is below the boiling point. The foodstuff thus cooked is easily digestible.

Positive nitrogen balance A condition where the intake of protein is higher than the output of protein. The growth process is identified with the positive nitrogen balance.

Poultry The meat of some birds comes under the category of poultry. Most poultry are purchased in the ready-to-cook form. The feather and viscera are removed and carcass is washed while dressing the poultry. It is also a nutritious food variety.

Proteins The principal nitrogenous constituents of protoplasm of all tissues of plants and animals. Proteins are necessary for the synthesis of body tissues and for many regulatory functions. The complex proteins are formed from the 21 amino acids in many different combinations. Proteins are otherwise called polypeptides.

Rancidity The phenomenon by which oils acquire foul smell on long exposure to atmospheric air and moisture. In this process, the unsaturated positions of the fatty acid chains are saturated with the peroxy-linkages.

Reducing sugars Sugars that react with Fehling's and Tollen's reagents and get oxidized by reducing the reagents.

Refrigeration and freezing The most popular food preservation technique. Refrigeration is done to slow down bacterial action so that food stays much longer (perhaps a week or two, rather than half a day) in unspoiled state. Freezing is done to stop bacterial action altogether.

Respiratory quotient The ratio between the carbon dioxide produced to the oxygen that is consumed in the respiration process.

Salting An ancient food preservation technique. The salt draws out moisture and creates an environment that is not suitable for bacterial growth. If salted in cold weather (so that the meat does not spoil while the salt has time to take effect), salted meat can last for years.

Saponification A process by which soap is formed by the reaction between oil and alkali. Soaps are sodium salts of fatty acids.

Saponification value The number of milligrams of KOH required to saponify one gram of fat or lipid. It is an index of mean molecular mass of the glycerides. It signifies the purity of the fat.

Seafoods Two major groups of fish are considered as seafood even though sea has innumerable living things. The two major groups are finfish and shellfish. The finfish variety includes whole, round, drawn, dressed, etc. The shellfish variety have soft body protected by shell— oyster, clam, scallop, etc.—and the segmented shell—lobster, shrimp, crab, etc.

Sieving A food processing method used to remove coarse fibres and insects.

Simmering Cooking the food by immersing in water at 80–90°C. While cooking, the vessel is covered with a lid. It is a mild and prolonged boiling method.

Simple lipids Esters of alcohol and fatty acids. For example, vegetable oils and some animal fats are simple lipids.

Simple proteins Proteins that contain only amino acids and their derivatives.

Simple sugars Sugars that are non-hydrolysable chemically and are otherwise called mono-saccharides.

Sodium A major nutrient useful in the regulation of acid–base balance and osmotic pressure in the body.

Sulphur A mineral nutrient useful in the formation of protein constituents.

Sweeteners Sweet-imparting substances. Table sugar is a sweet carbohydrate called sucrose. It is made up of two smaller sugar molecules (glucose and fructose) joined and so it is also known as a disaccharide. There are sweeteners that are not natural, but, artificial.

Terminal residue analysis The analysis of proteins which is done by chipping off one amino acid residue from one end of the polypeptide chain to know the exact sequencing of amino acids. There are two approaches— N-terminal analysis and C-terminal analysis.

Trace minerals The nutrients required in very small amounts, for example, iron, copper, iodine, manganese, cobalt, zinc, molybdenum.

Under-nutrition The deficiency of calories and/or deficiency of one or more nutrients.

Vitamin A The vitamin useful for the eyesight or vision. Mc Collum and Davis of Johns Hopkins University, Baltimore, identified it first. Vitamin A deficiency leads to night blindness. The diterpenoid unit of vitamin A is known as retinol or axerophthol.

Vitamin B A fat-soluble vitamin. In 1912, Funk found out that this substance was very effective in preventing beriberi. Mc Collum and Davis termed this vitamin as a water-soluble B. Vitamin B is not a single substance. It is a group of compounds, which we designate as vitamin B-complex. They possess very important biological functions as they form a part of various enzyme systems.

Vitamin B₁ (Thiamine) Otherwise termed as antiberiberi factor. It occurs in yeast, milk, groundnut, egg and outer cover of rice, wheat, etc.

Vitamin B₂ (Riboflavin) The yellow pigment of milk whey. In 1932 the German Chemist Warburg discovered vitamin B₂ as the yellow substance containing the pentose sugar to be present in the yeast. Thus, it was named Riboflavin (*ribo* = sugar; *flavus* = yellow).

Vitamin B₅ (Pantothenic acid) The vitamin isolated and synthesized by R. J. William in 1938. The Greek word *pantothen* means "in every corner or from all sides". Intestinal bacteria synthesize considerable amount of this compound.

Vitamin B₆ (Pyridoxine) Otherwise called antidermatitic factor; the name derived from its functionality.

Vitamin B₉ (Folic acid) Stokstad and Manning identified this vitamin in the year 1938 as a growth factor and termed vitamin U. Later, in 1945 Angier, Stokstad and their team synthesized this vitamin. The name folic acid was given due to its origin, that is, the dark green leaves of spinach (L. *folium* = leaf).

Vitamin B₁₂ (Cobalamine) The vitamin that can cure blood-forming defect and pernicious anaemia. It was labelled as vitamin B₁₂, in 1948 by two teams of scientists—one American and the other English—who crystallized it as a red compound from liver.

Vitamin C (Ascorbic acid)   As early as 1500 BC, Hippocrates, the "Father of Medicine", was aware of the disease scurvy. Jacques Cartier cured the dying men of his team, by giving a brew made of pine needles and bark. Norwegian scientists in 1907 reported that the food deficient in ascorbic acid could cause scurvy. In 1932 Charles Glen King and W. A. Waugh University of Pittsburg isolated hexuronic acid in lemon. They proved that it could cure scurvy.

Vitamin D   A group of five closely related fat-soluble vitamins—D_1, D_2, D_3, D_4 and D_5. They are structurally related to sterols. When sterols are irradiated with UV light, these vitamins are produced. Mellanby discovered vitamin D in the year 1919. Rickets is a deficiency disease indicated by the bulging of the forehead, wrists, knees and ankle joints and bone becomes soft and fragile.

Vitamin E   The vitamin necessary for reproduction. Evans and Bishop established the fact that a fat-soluble chemical factor was very much necessary for the reproduction in rats and they named it tocopherol. Of the two forms of vitamin E (tocopherol and tocorienol), the most active form of vitamin E is α-tocopherol. Reproductive failure is a deficiency disorder.

Vitamin K   It was identified as a *Koagulations* Vitamin by Professor Henrik Dam (Biochemistry, University of Copenhagen) in the year 1929. He found that chicks showed haemorrhagic disease when fed with fat-free diet due to the absence of a factor responsible for the clotting of blood. The haemorrhagic symptom is the deficiency disorder. The liver produces prothrombin if vitamin K is to be effective.

Vitamins   One of the nutrients of food. The term "vitamin" is derived from the term *vital-amine* as it was identified to be a nitrogen-containing compound that was vital for life. Vitamins are classified into two types based on the solubility in water. Some vitamins are soluble in water (water-soluble vitamins). Some other vitamins are soluble only in non-polar solvents (fat-soluble vitamins).

Water   An essential nutrient that must be supplied from an outside source because the body cannot make it in sufficient amounts. Without water, human life cannot survive. Water deprivation kills faster than lack of any other nutrient.

Water balance   The balance between water loss and water intake. The body has a sophisticated system that works to maintain water balance. Few of us ever experience malfunctioning of this system. Thirst is a trigger that reminds us to take in more water. At the same time our kidneys regulate urinary output.

Wheat flour   Flour obtained by grinding whole-wheat grains to a fine powder. The grinding or milling of wheat grain is done based on the grain structure. There are slight differences in the methods adopted depending on the grain variety.

REFERENCES

Bailey, J.L. *Techniques in Protein Chemistry,* 2nd edn. Elsevier Publishing Co., Amsterdam. 1967.

Dominic, W.S. Wong. *Mechanism and Theory in Food Chemistry.* CBS Publishers and Distributors, New Delhi. 1996.

Hart, F.L. and Fisher, H.J. *Modern Food Analysis.* Springer-Verlag, New York. 1971.

Harwood, L.M. and Mood, C.J. *Experimental Organic Chemistry.* Blackwell Scientific Publications, Oxford, London. 1989.

Joslyn, M.A. *Methods of Food Analysis,* 2nd edn. Academic Press, New York. 1970.

Leonard, W. Aurand, Edwin Woods, A. and Marion, R. Wells. *Food Composition and Analysis.* van Nostrand Reinhold Co. Ltd., New York. 1987.

Lineback, D.R. and Inglett, G.E. *Food Carbohydrates.* AVI Publishing Co., Westport, CT. 1982.

Pearson, D. *Laboratory Techniques in Food Analysis.* John Wiley and Sons, New York. 1973.

Pomerang, T. (ed.). *Wheat Chemistry and Technology,* 2nd edn. American Association of Cereals Chemists, St. Paul, MN. 1971.

Raj, K. Bansal. *Laboratory Manual of Organic Chemistry.* New Age International (P) Ltd., New Delhi. 1996.

Ranganna, S. *Manual Analysis of Fruits and Vegetable Products.* McGraw-Hill, New Delhi. 1976.

Roberts, R.M., Rodewald, L.B. and Wingrov, A.S., *An Introduction to Modern Experimental Organic Chemistry.* Holt Rienhart and Winston, New York, 1983.

Sonntag, N.O.V. "Structure and composition of fats and oils." In: *Bailey's Industrial Oil and Fat Products.* Vol. I. John Wiley and Sons, New York. 1979.

Sue Rodwell Williams. *Nutrition and Diet Therapy,* 3rd edn. C V Mosby Co., Saint Louis. 1977.

Vogel, A.I. *Qualitative Organic Analysis.* Longman (ELBS), London. 1972.

Webb, B. and Johnson, A.H. *Fundamentals of Dairy Chemistry.* AVI Publishing Co., Westport, CT. 1975.

Woods, A.E. and Aurand, L.W. *Laboratory Manual of Food Chemistry.* AVI Publishing Co., Westport, CT. 1977.

Zapsalis, C. and Beck, A.R. *Food Chemistry and Nutritional Biochemistry.* John Wiley and Sons, New York. 1985.

INDEX